The MILITARY and MODERNIZATION

The MILITARY and MODERNIZATION

Henry Bienen
editor
with a new preface by the editor

AldineTransaction
A Division of Transaction Publishers
New Brunswick (U.S.A.) and London (U.K.)

This book is printed on acid-free paper that meets the American National Standard for Permanence of Paper for Printed Library Materials.

Library of Congress Catalog Number: 2008027959
ISBN: 978-0-202-36305-9
Printed in the United States of America

Library of Congress Cataloging-in-Publication Data

Bienen, Henry.
The military and modernization / Henry Bienen.
p. cm.
Includes bibliographical references and index.
ISBN 978-0-202-36305-9 (alk. paper)
1. Armed Forces--Political activity. 2. Civil-military relations. 3. Political science. I. Title.

UA15.B5 2008
322'.5091724--dc22

2008027959

to my father,
Mitchell R. Bienen

Contents

Preface to the AldineTransaction Edition ix
Preface xiii

The Background to Contemporary Study of Militaries and Modernization 1
HENRY BIENEN

Further Reading 35

1 : *Armed Force Organizations* 41
MARION J. LEVY, JR.

2 : *Political Development and Military Intervention in Latin America* 79
MARTIN C. NEEDLER

3 : *The Military and Political Change in Asia* 102
JOHN P. LOVELL and C. I. EUGENE KIM

4 : *Swords and Ploughshares: The Turkish Army as a Modernizing Force* 117
DANIEL LERNER and RICHARD D. ROBINSON

5 : *The Army, Politics, and the People* 149

A. ISKENDEROV

6 : *Political Development and Political Decay* 157

SAMUEL P. HUNTINGTON

7 : *The African Military and Political Development* 212

CLAUDE E. WELCH, JR.

Index 235

Preface to the AldineTransaction Edition

The Military and Modernization was published in 1971. As I noted in the original preface, the book was part of a growing literature on the military in developing countries. Works on militaries examined armed forces' interventions in politics from coups to takeovers. The literature also analyzed militaries as rulers.

Analyses of military rule in developing countries involved discussion of armed forces as institutions. In turn, this meant looking at the recruitment, social composition, organization and functioning of militaries. It involved, also, seeing whether it made any difference for social, political and economic outcomes if militaries ruled. Did they perform bureaucratic and political roles differently from civilians?

Perhaps needless to say, there were differences between countries and within and between regions. History and culture mattered. Size of country and armed forces; levels of economic development, variations in ethnic heterogeneity, were consequential for outcomes and the literature explored these factors.

In the more than thirty-five years since *The Military and Modernization* appeared, armed forces in developing countries have evolved. Inter-country wars occurred: India and Pakistan; China and Vietnam; wars between Israel and various Arab countries; the United States and Vietnam; Armenia and Azerbaijan; the United States and Iraq; Argentina and the United Kingdom; Ethiopia and Eritrea. This is by no means an exhaustive list.

Countries dissolved or parts of countries split off, bloody civil wars sometimes became interstate wars: the creation of Bangladesh; Eritrea and then Ethiopia-Eritrea wars; the fragmentation of the former Yugoslavia and the Balkan wars are all examples.

Civil war and the weakening of states led to invasions by outside forces: the Soviet Union and Afghanistan; Rwanda, Angola, and Uganda in Zaire; South Africa, Zaire and Cuba intervened in Angola. Ethnic and territorial conflict most recently opened this door for the Russian invasion of Georgia and of Uganda and Rwandan exiles into Rwanda. Again, these lists are by no means exhaustive.

Insurgencies have abounded and militia continue to proliferate sometimes forming the armed basis for a state within a state; sometimes simply being able to create instability. Hezbollah in Lebanon; various militias in Georgia; the Sudan; East Timor; Zaire; Somali Democratic Republic; Uganda; Iraq are examples.

And, armed forces continue to carry out coups or attempted coups and have been the arbiters of regime change. This occurred most recently in Mauritania. It has occurred in many states since *The Military and Modernization* was published; among them: Venezuela; Ivory Coast; Turkey; Argentina; Madagascar; Liberia; Sierra Leone; Thailand; Bolivia; Bangladesh; Pakistan; Equatorial Guinea; Fiji; Niger; Nigeria; Benin; Sudan; Uganda; Ethiopia; Guinea Bissau; Rwanda; Portugal. There have been many unsuccessful coups too. Among them Ecuador, Guinea-Bissau, Central African Republic, Benin, Bangladesh, Mali, Nigeria, and on and on.

The obvious point of these non exhaustive lists is to note that the role of armed forces in developing countries has not declined over the past decades. It would be hard to find neat patterns for military involvement. Outright takeovers continued to occur as did coups and counter-coups. Military rule could be short lived or long run. Militaries ruled with varying degree of civilian involvement.

I have not tried to correlate levels of economic development or degrees of ethnic heterogeneity with types of military involvement in politics since 1971, although studies have carried out these analyses. Nor have I tried to examine the impact of military rule on economic performance or political stability. Indeed, this new preface is not by way of introducing new work so much as it calls attention to the continued relevance of armed forces both as an independent factor and as society works in on military institutions.

Perhaps, if I were setting out now to edit a volume of new essays on armed forces I would not title such a work, "The Military and Modernization." I would instead title it, "Armed Forces and Society" in order to try to call attention to the many ways militaries both grow out of societies and effect them.

Others can judge whether these essays still shed light on important issues, whether they foreshadowed work that came after them, and whether they at all still point to work that needs to be carried out.

Henry Bienen
Evanston, IL
September 2008

Preface

When I first agreed to contribute to the Atherton *Controversy* series in 1967 there was already a growing body of work on the military in developing countries. Few studies, however, explored in depth questions concerning the social origins of officers and enlisted men or traced career patterns within the armed forces of a specific Asian, Middle Eastern, African, or Latin American country. With the exception of Latin America, it was a rare study that assessed the performance of ruling, or nonruling, militaries for political development and modernization of their societies. Most studies analyzed the coup process and speculated in a general way on armed forces and modernizing institutions.

In the intervening years, interrupted by my stay in Kenya, the literature on militaries in developing countries has both widened and deepened. We are beginning to get the hard information we need in order to come to grips with the propositions that have already been formulated. We need no longer treat militaries in non-Western countries in a relatively undifferentiated fashion. It has been in part due to the efforts of the Inter-University Seminar on Armed Forces and Society that progress has been made in the study of armed forces organizations. My own thinking about and writing on the military has in the past received Inter-University Seminar support and I take this occasion to acknowledge it gratefully. I want also to voice my thanks to the Russell Sage Foundation, whose financial assistance has made possible important research in this field. The Chairman of the Inter-University Seminar, Professor Morris Janowitz, has himself set high standards in the study of armed forces organizations. I am grateful to Professor N. Wahl for his thoughtful comments and helpful editorial work on this manuscript.

Finally, thanks are due to the Center of International Studies at Princeton University, which has given me general support for research and great help in preparation of manuscripts. The efficient and cheerful secretarial staff of the Center of International Studies has my gratitude.

The Background to Contemporary Study of Militaries and Modernization

HENRY BIENEN

Warfare, violence, and armies have always been a matter of interest to political analysts and practitioners, but American social scientists in increasing numbers concerned themselves with civilian-military relations after World War II.[1] In part, this was a reflection of the recent experiences of large numbers of them who had been in the military. It also reflected a new awareness of the importance of armies in peacetime as well as in war. Where civilians ruled, in the United States and most of Western Europe, problems to be investigated included the military's organizational structure and social composition, its role as an interest group in foreign and domestic policy making, and the impact of the military on domestic politics and economy. Indeed, the concern with military-civilian relations in the Western context reflected and often echoed the old European debates about militarism. Is the army a school for political virtue or a retrograde force in society? Are soldiers still to be considered heroic but violent, the traditional point of view, or are they new bureaucrats with highly refined technical skills transferable to civilian spheres? Are the armed forces of Western Europe and the United States the carriers of a new militarism or the most advanced example of rationalized organization involved in problem solving?[2]

A controversy later emerged over the same themes applied to non-Western countries. However, in the context of developing countries, it was above all the role of the military as a modernizing force in society around which debate centered. Of course, implicit in the work on post-World War II European and American armed forces was concern with modernization, since what was at issue was the armed forces' abilities to promote new technology and organize economic and political power. But the emphasis was more on organizational interest group politics in industrialized society than on institutions for modernizing those societies. (The nineteenth-century French and German writers on militaries and contemporary historians looking back at France and Prussia were more explicitly concerned with modernization problems.)

Modernization and political development themselves became "issues" through concern with new states. Once attention was focused on the study of developing areas it was necessary to consider the military as the maker of coups and the ruler of states and not merely as an important interest group in society. Ironically, the organizational structure of specific militaries and the actual social composition of given military forces in Latin America, Asia, and Africa were not established. Data was and is hard to come by in these areas. But this is only a partial explanation of this phenomenon, for the role of militaries as interest groups in developing areas has been neglected also, with the exception of Latin America. And there has been little evaluation of actual performance of ruling military regimes. Thus, although attention has been paid to military institutions since the 1960s, the debate over the role of the military as a modernizer in Asia, Africa, and Latin America has taken place in something of an empirical vacuum. It has been a debate to some extent uninformed by the kind of data gathering that took place in the study of militaries in the United States and Western Europe.[3]

This is not to say that nothing useful has been done on the military in developing areas. There have been a number of provocative essays on the military in new states[4] and fewer studies in detail and depth of the military in a particular country.[5] Most of the latter have been published as essays on the military of a geographic area.[6] There have also been selections devoted to the military in anthologies dealing with modernization or development in new nations. Finally, collections of essays on the military in non-Western areas have flowered.[7] The last is a sure sign of the explosion of interest on the subject.

It is this very proliferation of works devoted to the military and development that makes it timely to assess where we are in the study of the military in developing areas. Are the militaries basic to the modernizing process in developing countries? What organizational and personnel strengths and weaknesses do they bring to the various economic, social, and political problems of development?

Basic Themes: Military Intervention and Performance

Analysis of armed forces in developing countries has been organized around two themes. Chronologically, the first concern was to explain military intervention, particularly in the form of coups. Increasingly, attention has been shifting to analysis of the consequences of military intervention in the development process. Observers now try to evaluate the performance of the military as a ruling group; the militaries' capacity for maintaining stability, creating political institutions, and achieving economic development and national integration are being studied. The role of the military as a modernizing force in society is now being examined.

I have argued elsewhere that it is most fruitful to link the study of military interventions to the study of the military as a modernizing institution.[8] When we understand the reasons for military interventions and as we explore the links between the military and other political institutions and social groups, we can better assess the prospects for the future evolution of the armed forces in society and we can better judge their capacities to deal with specific problems.

Furthermore, the factors that explain military intervention also reveal the limits of the military's political capacity. Also, where coups come fast and furious, the process of making coups cannot be separated from a period of military rule.

If we can show that the military intervenes in a specific situation because it feels its own professional values are threatened and its corporate identity is at state, then we may be some distance toward explaining its subsequent actions as a ruling group. Similarly, if we can show that the military, or an element of it, makes a coup because it has a specific program for modernizing society that it feels only it can carry out, then we may be able to more fully understand military performance as a governing elite. Perhaps some of the controversy in analysis of the military as a modernizer stems from a failure to link rule and intervention empirically.

One of the merits of Martin C. Needler's "Political Development and Military Intervention in Latin America" (Chapter 2) is that Needler makes an argument about the consequences of military regimes in Latin America by examining the dynamics of coups d'état. He starts from the position that the militaries are not homogeneous organizations, a fact we sometimes forget when we speak about "the military." The political viewpoints of military groups can be seen by examining the internal dynamics of coups. Needler concludes from his study of coups that they are more likely to occur when economic conditions are deteriorating and that military intervention increasingly takes the form of an attempt by the possessing classes to maintain the status quo in Latin America. Even if Needler is correct on this point, we would not establish the proposition that militaries who hope to thwart social change and who succeed in seizing power necessarily accomplish their social aims. Huntington argues that "modernization is not the product of any one particular group, however 'modernized' that group may be in comparison with the rest of society. Rather it is the product of coup and countercoup in which military elements play important roles inaugurating both conservative and radical regimes."[9] Huntington's argument does not depend on the military being a comparatively modern organization in society; in fact, he states that the Latin American experience suggests that the militaries are neither exclusively harbingers of modernization nor the defenders of entrenched oligarchies.[10] In any case, conservative groups are not able to undo change, and radical coup-makers keep extending political participation and emphasizing reform.[11]

Needler too sees developmental trends. He notes that observers are divided into two groups, those who perceive the revolutionary forces at work in Latin America as tending in a democratic direction and those who say *"plus ça change. . . ."* Needler maintains that a cyclical pattern of dictatorship reproduces itself around a clearly descending line "so that each successive peak in the number of dictatorships existing contemporaneously is lower than the last." If we put this conclusion together with Needler's point that the purpose of coups is increasingly to thwart social and economic change we arrive at the position that social and economic development take place anyway, despite military intervention.

The question remains: in what ways does a ruling military function to further change, albeit sometimes despite itself? In order to come to grips with this question, it may be useful to think of three rough phases of military intervention which are defined by different political contexts, for just as we want to get away from the notion of *the* military within a given country, so we should want to differentiate situations in which militaries act.

There is an initial stage where the military may for the first time make overtly threatening demands on the government of a new nation; it may settle conflicts between civilian contenders or it may itself take over the government. Tropical Africa now presents us with examples of this phase. The next phase can be considered one in which there is a struggle for stability after the seizure of power. The criteria for inclusion in this category are: (1) a regime dominated by the armed forces, (2) a military leadership that has a great deal of continuity with the original leaders of the coup, (3) a situation that has not been characterized by a number of coups and countercoups. Thus we are still dealing here with a first generation of military leaders and a situation now found in such countries as Korea, Burma, Pakistan, and the United Arab Republic, although some change in the top leadership may take place.

The third of our rough categories is that of institutionalized intervention. Latin America presents us with a host of examples. Here the military cannot bring about political stability and it cannot give up aspirations to rule. Military coup may become the accepted way of getting political change for groups within the armed forces, for certain civilian leaders, and for citizens who remain bystanders to coups.[12] In each of the three phases, the military has different immediate political problems. In the first it must pick up the political pieces left by civilian regimes or by its own coup and must find allies and support from political groups. In the second it must build support, reach down into society, and maintain military cohesion. In the third phase the military is split, by definition, and must maintain some autonomy from civilian groups and keep military disputes within bounds. In all three phases in developing countries, the military invariably sets itself the task of providing political stability and public order and often tries to promote economic development. The compatibility of these goals may be difficult to achieve, as, for example, when higher taxes are called for and lower civil service salaries seem necessary. As Huntington points out, modernization and political stability may be even harder to make compatible as the former includes more than economic development per se.

In each phase of military rule, then, the military has different problems vis-à-vis its own internal politics and its confrontation with the society at large. Problems of modernization and development overlap our three categories because different levels of income and various social structures can be found in any and all of the categories we present here. Nonetheless, it makes sense to draw these distinctions as we confront questions about the military and development.

The Argument for the Military as a Modernizing Force in the Ascendance

It has been said that there is an American liberal tradition that makes it repugnant to contemplate regimes controlled by military elements rather than by civilian instrumentalities of representative government.[13] It has also been suggested that American antimilitarism includes a peculiar selectivity about America's own historical experiences since the American military was itself a nation-builder through civic action programs.[14] Academics in particular have been accused of a parochialism with regard to armed forces.[15]

These accusations may well have had force until the end of the 1950s. But if there has been an American antimilitarism forcibly expressed in American scholars' bias against the military, so that scholars were both inattentive to the sociology of armies and also refused to recognize national security problems as a respectable concern, it cannot be said that American governments consistently shared this bias or that academics still cling to it.[16]

Lucien Pye pointed out in the early 1960s that we were confronted by two conflicting images of the politician in uniform. The first, Pye said, was derived largely from Latin America and the Balkans and was one of "administrative incompetence, inaction, and authoritarian, if not reactionary, values." The more recent one was of a "dynamic and self-sacrificing military leadership committed to progress and the task of modernizing transitional societies that have been subverted by the 'corrupt practices' of politicians."[17] Of course, the very fact that Pye had referred to the "politician in uniform" marked a change in terminology as applied to the military. By the mid-1960s, the armed forces in new states were more often described as modernizing and nation-builders than as dictatorial and backward-looking both in the general literature on new states and in writing on the military as a political institution. One indicator of this was the prominence of Pye's own article on "Armies in the Process of Political Modernization."

At first, as is often the case with new departures from received wisdom, the reaction against the old line was a moderate one. Guy Pauker made the argument that without exception the countries of Southeast Asia were likely to experience increasing political instability. Their leaders had neither the prestige of traditional aristocrats nor the prestige of the first-generation nationalist leaders. Politicians would not have time to establish efficient democratic government and administration in a leisurely way because well-organized communist minorities would fill power vacuums in the context of general instability. "Those best equipped to become an effective counterbalance to the spread of communism in Southeast Asia are members of the national officer corps as individuals and the national armies as organizational structures." Pauker noted that some knowledgeable observers saw the officer corps of Southeast Asia as not different from the rest of their societies, possessing only the advantage of the control over means of coercion. But Pauker stressed that the officers were indeed new men. He did not spell out the organizational characteristics of military leaders in terms of Gemeinschaft-Gesellschaft dichotomies, but he did note their personal qualities of leadership, patriotism, and commitment to moral values and saw them as a social product of modernization and the only noncommunist group equipped to modernize their countries further. They had some experience in administration of civilian affairs, although their technical competence in economic and administrative matters was generally limited. Pauker concluded that the officer corps of Southeast Asia would be able to acquire political skills in the short time available before their societies disintegrated.

Pauker's argument was defensive in tone. He did not insist that rule by the military was the best of all worlds but only that regimes controlled by the military still left open the possibility of development in a democratic direction.[18] In any case, the only alternatives were communism or disintegration.

Also published in 1959 was the *Report of the President's Committee to Study the United States Military Assistance Program,* better known as the Draper Report. Taken together, these works constituted a major event in the analysis of the politics of new states. The Draper Report made a positive assessment of the capacities of militaries in non-Western countries to carry out nation-building programs. Armies existed, they would not go away, and they often were the only institution available to make headway in economic and social development programs.[19]

The pendulum continued to swing, and since 1959 those who have argued that the military is the organization best able to bring about development in non-Western countries have not been arguing a minority position. A bandwagon effect operated in the controversy over the role of the military in developing areas and academic opponents of the conception of militaries as modernizing institutions largely left the field between 1960 and 1965. During this period, insofar as doubt was cast on the role of the military as a nation-builder it came less from the universities than from Congress and the editorial pages of the *New York Times.*

It was in the Congressional hearings on military assistance that the value of direct military rule for economic growth and political stability were questioned.[20] Some Congressmen have been opposed to military assistance because they are opposed to United States involvement overseas; others are more concerned with ending foreign aid in general than with caring much about political development in Asia, Africa, and Latin America. In any case, the views of Congressional opponents of military assistance have had relatively little resonance among analysts of political development in recent years, with the exception of one area. Among students of Latin America there has been much more doubt and ambivalence about the role of the military. Observers of Latin American politics retained their skepticism about the military as a modernizing force. Some continued to see the military as a corporate, self-interested group with institutional conservatism that inhibited economic development and social reform. Others have argued that modernization in Latin America could not be brought about by reformist militaries but only by revolution and that regular armed forces were the principal obstacles to revolutionary change.[21]

A number of people who surveyed military-civilian relations in the developing countries pointed out the peculiarities of Latin America to those who wished to generalize about the Third World. Morris Janowitz warned that while Latin American countries were confronted with crises of civil-military relations that looked similar to those of the new nations of Asia and Africa, there were fundamental differences in the natural history of militarism in South America. Janowitz noted that although the comparative study of the military in politics required exploration of Latin American nations, the forms of military intervention represented "more than a century of struggle and accommodation which has produced political institutions different from those found in the new nations."[22]

José Nun has pointed to the greater degree of industrial expansion in South American countries, particularly in Argentina and Brazil, compared to developing countries elsewhere. He notes the less recent and direct colonial experience of Latin America as compared to Asia and Africa. Above all, Nun emphasizes the greater integration of major Latin American armed forces with their societies and the greater institutional complexity of those societies and militaries as compared to most developing countries elsewhere. Nun argues that military intervention in Latin America does not threaten the middle class nor is it a substitute for its absence but that the armed forces tend to represent that class and compensate for its inability to establish a well-integrated and hegemonic group. At the same time, Nun insists that while one can distinguish between "progressive" and "regressive" varieties of Caesarism, military intervention in its present form in Latin America tends to prevent the possibility that certain sectors of the middle class will transcend their "profoundly traditional" outlook. Thus, Nun believes, the military is not a progressive force in Latin America because the relationship of the middle classes to the lower classes inhibits social change and development.[23]

Nun's argument is an important one, and not only for Latin America. For while there was more doubt cast on the modernizing abilities of armed forces in Latin America than were raised about their counterparts in Asia, Africa, and the Middle East, even analysts of Latin America began to share a view of the military as a force for national unity and development through the use of technical abilities, identification with the lower middle strata of the population, and availability of political leadership.[24] True, it was still argued that military values tend to supersede civilian ones and that many young men become officers in rejection of liberal middle class values. (This is what Nun has called the "liberal model" of military-civilian relations, a model based on the experience of Europe in the eighteenth and nineteenth centuries which envisages the army as the bastion of traditional and feudal values.) But increasingly influential was the idea that recruitment of officers from middle and lower strata will produce important changes in military values and that younger officers will tend to retain the liberal and progressive values of their families.[25] Thus, as Nun questioned the progressive nature of the middle class in Latin America he went to the heart of the argument for the military as a modernizing institution because many of the arguments about the organization and efficiency of the military are related to the assumption that members of the military are recruited from a comparatively modern, rational, and cohesive middle class.

The Argument for the Military as a Modernizer in Its Pure and Other Forms

Not all arguments for the military as a modernizing force have been empirically derived. Marion J. Levy, Jr. has argued that armed forces organizations are the "most efficient type of organization for combining maximum rates of modernization with maximum levels of stability and control."[26] He has argued this position in its purest form, that is, unencumbered by empirical detail and at a high level of generality, in his paper "Armed Force Organizations" (Chapter 1 of this book). In this work, Levy looks at the internal characteristics of armed force organizations and places armed forces in their social setting. He notes that "professional soldiers and their subordinates have not been conspicuously successful in social history in long-run governmental administration conducted primarily in terms of armed force organizations." This is because "preoccupation about the rational application of [force and high levels of coercion] tends to preclude consideration of other measures. In the context of all viable armed force systems there is an emphasis on *command* which is never easily identifiable with an emphasis on *governing* in general." For Levy, the frequent emergence of armed force organizations and their leaders in central government positions in developing areas (which Levy calls "late-comers to modernization") is not a contradiction to his points cited above but is rather a function of the fact that armed force organizations are likely to be the foci of modernization and ladders of social mobility for talent in their societies. The armed forces may be the major setting for bureaucratic experience, and an increased proliferation of bureaucratic organizations is one of the inevitable changes in any modernizing society.

We should state that Levy defines modernization in the following way: "The greater the ratio of inanimate to animate sources of power and the greater the multiplication of effort as the effect of application of tools, the greater is the degree of modernization."[27] In relatively nonmodernized societies, Levy maintains, "the emphasis on modernization in general with regard to armed force organization is an emphasis which generates little resistance." Fundamentalist objections to change pose fewer problems with regard to the armed forces. Universalistic criteria based on physical fitness and mental ability will be more predominant in recruitment to armed force organizations than elsewhere; scale of size and recruitment procedures make armed force organizations perhaps the major ladder of social mobility; the military harbors radical feelings about the status of the nation as a whole; there is a minimization of the problem of feedback to the population as a whole because contact can be channeled between members and nonmembers of armed force organizations.

In Levy's analysis of modernization, the impact of ideas and institutions developed in relatively modern societies on relatively nonmodern ones is stressed. Lovell and Kim develop this theme in their work on Asian militaries (see Chapter 3). They assert that the military is a key mechanism for receiving and sometimes amplifying signals from the external environment. Militaries, it is said, have a professional language that links them to other military forces abroad. Because relationships with the United States have been important, Asian military men in particular have had to develop a facility for cross-cultural professional language.

Lovell and Kim also emphasize the critical communications and socialization functions that armed forces perform through contacts with civilians and intramilitary networks. The military not only serves as a model but also conspicuously indoctrinates and trains. Lovell and Kim underline the importance of military activities for both intrasocietal and international communications in Asia. Here a number of the points that Levy states at a high level of generality are supported by concrete references to particular Asian armed forces. At the same time, Lovell and Kim recognize that we lack adequate information about military activities and their consequences which would allow us to formulate hypotheses for many important areas of analysis. Without having recourse to a large amount of information about specific militaries, Levy is prepared to argue that the military is the organization best suited to bring about modernization in relatively nonmodernized societies. He does not, however, directly deal with the question of whether or not the military must rule relatively nonmodernized societies in order to maximize modernizing tendencies.

Lucian Pye does take up this question to some extent as he

moves from consideration of the army as one of the more modernized of the authoritative agencies of government in what he calls transitional societies to consider it as a modernizing force for the whole of society. Pye outlines three general categories of relationships between military and civilian leaders. In one, members of the military stand out in a disrupted society "because they represent the only effective organized element capable of competing for political power and formulating public policy." In a second category, the military formally espouses the development of democracy but actually monopolizes the political arena and forces civilian elites to concentrate on economic and social activities. The third case is one in which "the process of modernization has been retarded to such a point that the army, as the most modernized organization in the society, has assumed an administrative role and taken over the country." Pye also warns that while armies may be an element of stability in underdeveloped countries, we should not confuse this with political stability for an entire society. The objective must be stable representative institutions and practices, and to achieve these more is involved than the efficient administration of public policies. For Pye, skilled and responsible politicians are necessary too, and he does not cast the military in this guise when he says they can contribute to only a limited part of national development.[28] However, implicit in Pye's analysis is the possibility of the military taking up political as well as administrative roles. As we noted earlier, he refers to "politicians in uniform." Pye does not tell us whether the politicians in uniform can best become skilled politicians by taking up mufti or remaining in khaki. What is at issue here is an evaluation of the Ataturk model for political development.

The Ataturk Model

Those who have argued most strongly for the military as a modernizing force in developing areas have not distinguished sharply between military regimes and civilian regimes with former military personnel in commanding positions. Rather, it has been those commentators who are highly aware of the limitations of military rule who have stressed the appropriateness of the Ataturk model for development. During the period in the 1960s when emphasis was placed on the virtues of the military in non-Western countries, Morris Janowitz continued to argue that it is easier for the military to accumulate power than to govern as a ruling group.[29] This point was made with varying degrees of clarity by many observers. Janowitz, however, spelled out the reasons for this in terms of the characteristics of military organizations. He argued that while the militaries of new nations are more politicized than their Western counterparts and may produce a cadre of political activists, the military does not have the appropriate political ideologies for rule. "While they are strongly nationalistic, oriented to collective and governmental enterprise, military officers are skeptical and even hostile to organized politics, especially as practiced by partisan groups." Janowitz maintained that pressure toward political involvement did not necessarily imply the development of skills appropriate for ruling. He argued that Turkey under Ataturk was the one case in which a military oligarchy made fundamental contributions to social and economic modernization. This was possible because Ataturk himself understood the limits of the military in politics and insisted that army officers who wanted to become directly involved in partisan politics leave the armed forces. This allowed the military to maintain its cohesion. But as Janowitz pointed out, Turkey's process of modernization was rooted in gradual change.[30] Moreover, few underdeveloped countries are under-populated and culturally homogeneous, have a peasant land-owning class isolated from urban politics, and are united against foreign enemies—conditions obtaining in Turkey after Ataturk came to power.[31]

Chapter 4, Lerner and Robinson's study of the Turkish army as a modernizing force, describes the attempt to maintain civilian supremacy and at the same time use the Turkish army for nation-building.[32] Lerner and Robinson hypothesize that civilian supremacy was maintained in Turkey because the civilian sector had been relatively successful and the military had nothing better to offer. Moreover, there was a mystique of civilian supremacy and a representation of military interests at the highest levels of government. Above all, they argue, Turkey underwent a speed-up of the tempo of social change but kept social stability. Ataturk, however, did not use the Turkish army as a major innovating force. Rather, according to Lerner and Robinson, he perceived that the Turkish army was not yet a modern force and that a modern military organization was impossible without modern science and industry, which in turn would be possible only after a transformation of the social and political system. This is an important point because it raises the question of how military organizations themselves become modernized.

Those who argue that the military is suited to undertake economic and social reform and that it provides a vehicle for national integration as it assimilates groups begin with the proposition that the military is at least relatively the most modernized group in society because its own requirements demand that it be rational, universalistic, and industry oriented; furthermore, it must have contacts with modern forces outside its own society and utilize an advanced technology. They do not show how the military gets to be this way. Ataturk turned the problem around, according to Lerner and Robinson, by making the modernization of the army dependent on the transformation of Turkish society.[33] In Turkey, the portion of the general budget allocated to national defense fell precipitously between 1926 and the early 1930s and did not rise to pre-1926 levels until 1939. This too is an important fact. The nature of the military as an instrument of change varies not only with its place in society as ruler or interest group but also with its own size and structure. If the military becomes very large compared to civilian elites, it may be taking such a large share of national income that economic development will be precluded. The military may well lose political effectiveness as it acquires the machinery of force. It is at least conceivable that military success in bringing about economic development and social change will vary proportionately with its success in restricting its own growth and importation of military hardware.[34] Manfred Halpern has argued that in the Middle East the capacity of armies for making effective use of men, machines, and organization has been warped by conspicuous consumption and continued dependence on outside sources of military assistance.[35]

This point has relevance for policies of military assistance. Moreover, if a military turns out not to be the organization best suited for modernizing its society, then it will be an organization particularly hard to get rid of as it achieves greater coercive strength through assistance. Fred Riggs has put it this way: "Thus the very conditions which bring armed forces into power are the imbalances in which stability and control have been undermined. Nor can an armed force, as a ruling circle, readily create the conditions for effective re-establishment of modernization since the price that must be paid is the conquest of power by new organizations which are not armed. Those who exercise irresponsible power are almost always reluctant to surrender control and become responsible public servants."[36]

The Pendulum Begins to Swing Again: Views of Nonmodernizing Militaries

Although there are as yet few studies in depth on the sociology of specific armed forces in developing areas, we have begun to get information that casts doubt on the proposition that the military is relatively the "most modern" organization in certain underdeveloped countries. Studies that focus on the problem of cohesion in the military itself show that some segments of armed forces, defined in terms of generation, clique, or armed service branch, may indeed exhibit the organizational conservatism that used to be attributed to military organizations per se in the older literature in European armed forces.[37]

Yet it is clear that armies are "heavy institutions" even where they barely meet the requirement of being an autonomous armed force and whether they are "modern" or not. The small armies of tropical Africa, which often have no more than two battalions of infantry troops with little logistical support and modest firepower, can intervene and overthrow civilian governments. African armies have small, indigenous officer corps often composed of former noncommissioned officers who are not more advanced in terms of education, absorption of rational norms, and exposure to outside influence than many civilians.[38] These armies have not proved to be disciplined forces as they intervene to bring down governments, go on pay strikes, and act without regard to chain of command orders even when no coup is attempted.

Thus, it is one thing to be disciplined enough to exert the force or threat of force required to bring down civilian regimes —even so-called strong one-party regimes—and it is another thing to be a disciplined instrument in the hands of commanding officers. While African soldiers were early carriers of nationalism, this does not mean that African armies always are a force for national unity. As McWilliams points out, African armies were often recruited from minority social strata or ethnic groups. In fact, they were often deliberately recruited from minority groups thought to be suspicious of nationalist movements dominated by an ethnic community like the Kikuyu in Kenya.[39] The tribal complexity of armies has been a political issue in Uganda, Dahomey, Nigeria, Kenya, and other African countries.

One would be hard put to argue that the failure of military regimes is unique in new states; the failure of modernization and political development is characteristic of many underdeveloped countries which have civilian regimes. But in the Nigerian case, as an extreme, the escalation of violence can be traced as a specific consequence of military coup and rule. These points made with reference to ethnic divisions can also be made with regard to class. As Halpern points out, the "middle class" is a wide umbrella: "Differences in political choices among members of the new middle class, however, also reflect differences in the structure of that class and the variant character of its class consciousness."[40] This is as true within the army as it is between the middle class–army officer corps and civilian middle class people.

It is not quite correct to say that the argument for the military as a modernizing force rests on the proposition that military groups have more "modern attitudes" and are more "advanced" than civilian elements.[41] This can be shown to be the case in some states, but it does not follow that the military is best suited to rule. Halpern and McWilliams both point to the Middle East as an area where it has been largely true. Armies entered politics in the Middle East after Arab defeats in 1948 because they knew themselves to be "more disciplined, coordinated, technologically skilled, and morally dedicated than the rest of society." Yet Arab armies continue to struggle against forces inside their societies that are in turn struggling for, or against, more rapid political, economic, and social transformation.[42] A great deal of effort is spent staying in control. Even in the midst of the most recent Arab-Israeli war, the armies would not take internal security for granted, and this apparently affected movement of Syrian and Jordanian units.

Armed forces can be more concerned with control than with modernization, and not only because they sometimes have a passion for control in which organization and routine replaces politics and personality.[43] The military may be concerned with control when it fails to gain consent as a ruling group.

It has been argued that African politics already manifest a major transformation, the shift away from political power as a technique of rule to reliance on force.[44] Rulers resort to force when power and legitimacy fail. The increasing intervention of African armies in politics is in good part a result of the reliance of civilian leaders on force, and thus there is an increase in the saliency of the instruments of force. If armies fail to legitimize themselves they have to rely on themselves as instruments of force rather than as manipulators of political power. Militaries as ruling groups may use the instrument of force more readily than civilians because that instrument is familiar and at hand and because of a disposition not to tolerate disorder and overt political opposition. We do not have enough information on military dispositions to state these propositions with a high degree of certainty. We can with perhaps more certainty continue to explore the limits on military rule by examining arguments for parties rather than militaries as modernizers of underdeveloped countries.

We have learned that the coups that bring the military to power often carry the seeds of factionalism within the military, if, indeed, the military is not split already.[45] If the military is to rule, it cannot stand above politics but it must be a political actor. Thus, a major claim for the military—that it represents the entire nation and is perceived by citizens to be above political strife—ceases to be viable once the military gains power. How much political talent and legitimacy specific armed forces have is a matter for investigation in each case. What we can perhaps generalize about is the ability of the military to get down to the grass roots. If the military cannot do this, then by definition rule is limited. Furthermore, information may not be forthcoming to the centers of military command, so that the military confronts a situation in which information and coercion are related inversely in the system.[46]

The question then is: What are the shared organizational characteristics that make militaries more or less likely to reach down to the grass roots, to mobilize people, and to direct social and economic change? This question can be complicated by asking about types of militaries and types of political and economic situations. But our immediate task is to examine an idea of political development that stresses the role of party rather than the military.

As I have noted, between 1959 and 1965 the predominant view of the role of the military in developing areas emphasized characteristics of armed forces organization which were functional for modernization. Those who shared this view saw the military as a relatively effective ruling group. During this time, there were, of course, varying degrees of commitment to the military on the part of scholars and policy makers. Military rule was seen as a practicable and stable alternative when parliamentary, democratic regimes faltered;[47] as the only noncommunist rule possible in power vacuums;[48] and as the most efficient organization for modernity and stability.[49] Throughout this period, however, the Ataturk model was held aloft by those who looked to other political organizations to bring about development. In this model, retired officers would augment civilian institutions.

With the publication of Samuel P. Huntington's essay "Political Development and Political Decay" in 1965 (see Chapter 6 of this volume), a frontal theoretical assault was made on the idea of the military as the best institution to bring about development in underdeveloped countries. But even prior to Huntington's article, those who espoused the Ataturk model had implied that civilian institutions, not military ones, were better suited to lead their countries to the promised land of modernized society. The argument for parties rather than civil services was made forcefully by Halpern, who had himself stressed the relative modernity of armed forces and their special political virtues in the Middle East. Halpern argued that armies leave civilians outside their ranks and that only parties were capable of instilling a sense of citizenship and organizing public participation in political decisions in the Middle East. Halpern maintained that only a party could be in daily contact with the constituency and teach, propagandize, or put pressure upon that constituency to adopt new ideas and patterns of action. Only a party could stimulate involvement and gather new talents and thus regularize recruitment into the new elite. Parties, since they were unknown in traditional Islamic society, could move beyond the established order. "Only parties can link leaders and masses in almost daily contact. . . . Only parties can organize enthusiasm on the basis of a solidarity of interests with citizens outside the government."[50]

Halpern emphasized mobilization and participation in development; McWilliams opted for parties rather than armies on the grounds that there is a dominant economic base to all development and thus the army is, in normal circumstances, a less desirable vanguard than the party. McWilliams argued that Lenin's model of a vanguard party is better adapted to states that are "retarded" politically but have "advanced" economic organizations in at least some sections of the economy. An army vanguard is appropriate only to states that are backward in both political and economic respects. And even then, McWilliams insisted, the success of the vanguard army demands that it transfer authority to civilian leaders because the army conceives of the nation as a backward form of itself; insofar as the army is successful, the nation will come to resemble it. The more complex the civilian society, the more dependent the military is on civilians who possess central managerial and technical skills. Thus, the more successful the army in modernizing, the more it modernizes itself out of its role as ruler.[51] This argument must face the work of Janowitz and others who see a "civilianizing" of the military in the United States as the military acquires precisely these managerial and technical skills.

A. Iskenderov, in his essay "The Army, Politics, and the People" (Chapter 5 of this volume), also makes an argument that stresses the changing role of the army as class situations change and as the strength of a revolutionary party grows. For Iskenderov, a Russian political analyst, the army can emerge as a kind of holding institution until the revolutionary party gathers strength. A revolutionary situation can arise and develop without an advanced political party, but only this kind of party can consolidate the victory of the revolution. In his analysis, the army cannot take the place of party. The revolutionary party must be the ultimate guide in building the new society. Iskenderov also argues that it will be a party which will mobilize the masses and bring them to full political participation. Iskenderov's analysis is interesting because it differentiates among armies in ex-colonial countries and looks at the relationships between changing social and economic conditions and changing military ideologies, recruitment, and roles.

Of all these arguments, Huntington's has been the most influential. He says that in the absence of traditional political institutions parties are the only modern organization that can become a source of effectively institutionalized authority.[52] In order to understand Huntington's analysis it must be made clear that he differentiates the concepts of modernization and political development, as many other commentators do not. For Huntington, political development is institutionalization, a process independent of, although affected by, the process of modernization. Huntington's idea of political development has to do with the strength and stability of political institutions. Political development can be measured by the scope of support for and the level of institutionalization of political organizations, and there are, in turn, measures for these. Thus, Huntington has cut political development away from ideas of mobilization and participation. He argues that rapid increases in mobilization and participation undermine political institutions, yet mobilization and participation are the principal political aspects of modernization. In Huntington's formulation, rapid modernization produces political decay, not political development.

Once having defined political development as the institutionalization of political organizations and procedures, Huntington proceeds to argue that military juntas or dictatorships confront a distinct set of problems in the conflict between their own impulses to modernization (economic development, social mobilization, social reform) and the needs of political institution-building. Because he accepts the view that the militaries in new states have absorbed rational norms and recognize the necessity of social reform, he sees the officers as modernizers par excellence. At the same time, however, they are "frequently indifferent or hostile to the needs of political institution-building. . . . They can eliminate or exclude from politics individual civilian politicians, but they are ill-prepared to make fundamental changes in political processes and institutions."[53]

For Huntington it is the political party, not the military, which can be effectively institutionalized as a ruling force: "The importance of the political party in providing legitimacy and stability in a modernizing political system varies inversely with the institutional inheritance of the system from traditional society. . . . Military coups and military juntas may spur modernization, but they cannot produce a stable political order." Thus, Huntington takes issue with those who see the military as the creator of development on their own ground; the problem of stability and order is paramount for him.

Huntington questions the Ataturk model when he says that if members of the military attempt to restore civilian government and to continue in power as a civilian political group, they open themselves to the influences that corrupted civilians in the first place and thus pave the way for a second military takeover by a younger generation. This has happened in Korea.[54] An analyst of Turkey in the 1960s argues that political activists in the Turkish military were unable to use military background and high position to fashion an effective political base for themselves. They were prevented by the presence of alternative and legitimate political organizations and by military superiors in the junta who demanded a return to civilian rule without personnel infusions from the military activists.[55]

In Chapter 7, Claude E. Welch applies some of Huntington's arguments to an analysis of African military regimes. His concern is to measure political capacity. Welch stresses the importance of nationalism in building "legitimate organizational strength" and in mobilizing sufficient resources to carry out tasks. He too looks at the Ataturk model and argues that the Turkish example is inapplicable to contemporary Africa because Turkey after the establishment of the Republic was a more homogenous community than are African states. He further argues that the Ataturk regime cannot be called a military one. Welch's prognosis is for political decay, not political development, in African states falling under military control.

Conclusions and Some Policy Implications

Few controversies in the politics of developing areas are as immediately connected to policy conclusions as that of the role of the military in the political, social, and economic development of non-Western countries. Pauker's aforementioned article was clearly written with the intent to persuade his audience of the need to support Southeast Asian militaries through diplomatic support and military assistance. Lerner and Robinson state that American military aid provided a stimulus for the revival of the Turkish military corps, which had become frozen and without its innovating character by 1948. "With American military aid came a new mission for the Turkish army, and the new concepts and methods needed to fulfill this mission." The Turkish army had to modernize its rural conscripts before it could become a modern fighting force. Lerner and Robinson write that Turks acquired literacy, new habits of dress, and a new personality; they achieved physical, social, and physic mobility through military training programs. The military became a major agency of social change as it spread new skills, machines, and concepts into the hinterland. Individuals who led the army, recruits as well as officers, became the driving force for technical and social innovation in the villages. Lerner and Robinson argue that American military assistance was essential for infusing the Turkish army with new spirit and organization as specific American military assistance missions trained Turks to carry out their own civic action programs. Lovell and Kim discuss the importance of civic action programs in Burma, the Philippines, and Korea, and the role of the United States assistance in the latter's endeavors. They also note the utilization of civic action techniques in communist regimes.

Insofar as militaries directly carry out nation-building tasks, they do so, in current terminology, through civic action programs. Civic action, however, stresses the nonviolent use of the military. "Civic action is the use of indigenous foreign military and paramilitary forces on projects useful to local populations in fields such as education, public works, agricultural, transportation, and other projects which contribute to economic and social improvement."[56]

Military Assistance Program (MAP) supported civic action is designed to encourage and support usage of local forces in activities that contribute to economic and social development and to assist in the prevention and elimination of insurgencies inimical to free world interest by improving the relations between army and population. Congress amended the Mutual Security Act in 1959 to provide specific legislative endorsement for civic action, and in 1961 President Kennedy directed the Departments of State and Defense to undertake expanded support of civic action in developing countries. Civic action also served as a way of getting economic aid into developing areas, and some programs were financed through the use of counterpart funds as well.

Within the State and Defense Departments, as interest quickened in the military as a force for development, civic action programs were seen as direct contributions to economic development and as useful components of counterinsurgency programs. The economic aspects of development were stressed. "The purpose of the nonmilitary operations, such as civic action, is to help eliminate the economic cause of discontent that provides the breeding ground for insurgency."[57] Civic action has often been treated as a subcategory of counterinsurgency, and the attention paid to it within government and by many academics stems from its relevance to insurgency and counterinsurgency. As Edward Glick points out, however, neither as a concept nor as a process need civic action depend upon insurgency and counterinsurgency.[58] It has to do with the social and economic activities of the military per se and is related to arguments about the role of the military in developing areas within a broader context than that of internal war.

Analysts of civic action programs examine the activities of both ruling and nonruling militaries. In the latter case, there has been less controversy over the role of the military as it undertakes civic action. Yet even here there has been a gradation of enthusiasm for civic action programs in developing areas, and disagreement has been evident over the role of United States military assistance The Draper Report makes a positive evaluation of the contribution of military resources to economic and social progress. This argument has been echoed and modified since the Report's publication. Many policy makers and scholars hold the view that militaries are going to play political roles no matter what position the United States takes and that militaries should therefore be trained to perform them effectively.[59] Civic action has been seen as a means for "subverting" military institutions by training them to peaceful uses. It has also been maintained that unless the military carries out civic action programs the necessary roads, schools, and hospitals will not be built.[60] There have been fewer negative statements about civic action carried out by ruling militaries although there has been much controversy about military assistance to developing countries ruled by civilians on the grounds that this will lead to military intervention and inflate military budgets.[61] However, there has been less attention paid to the argument that even if coups will come about without military assistance, that the United States may have relatively little influence in bringing about coups, still, military assistance can affect political change, and civic action may itself be dysfunctional for political development.[62]

There are some obvious problems with civic action programs. They may raise expectations that cannot be fulfilled; they may be one-shot forays of the military into a rural area; they may compete with civilian programs for men and money; they may antagonize people if imposed on the countryside; they may be fragmentary and not fit into an overall social and economic policy for development. But the crux of any argument with regard to civic action programs has to do with the military and political development. Those who propound civic action as a strategy for development usually do not have in mind the United States example of the military's important role in "opening up the West," although American experience is widely cited. Rather, the military is seen as a surrogate for a civilian government that will not or cannot engage in social and economic reform and cannot hold the fort against insurgency, whether the military takes over or not. This brings us right back to political issues. Is the military itself reform minded? Can it sink roots into the rural areas, which it must do if it is to have a continuous impact? What kind of national and local alliances will military groups form? Have armies the organization, composition, and values suitable for social and economic reform that is dependent on political change?

Civic action, then, cannot be considered apart from the role of the military in the politics of development in Asia, Africa, and Latin America. Those who say the militaries are on the scene and will remain so in important ways are correct. This does not make it any the less incumbent on us to analyze the military in developing areas.

We are now at the point where important theoretical arguments against the position that the military is either an ideal or relatively good ruling organization to bring modernization to developing areas are being heard. And a number of case studies of military regime performance are casting doubt on some of the statements of the early 1960s. There has not yet emerged from the controversy as it is reproduced in the pages that follow any clear synthesis of positions or arguments that have been dominant for many years. As I have shown there has been a certain chronological shift, but at no time has consensus been complete on the role of the military in development. Nor is any overarching position likely to be couched in pro or con terms on this issue. The more hard data we get on the sociology of armed forces and the more informed the studies that emerge on the actual performance of militaries in power, the more complex the arguments should become. We should be able to understand better the possibilities and constraints of military rule and functioning in developing countries as we understand better the nature of militaries and the societies within which they operate.

As we suggested earlier, analysis of coups and analysis of subsequent military regime performance must go hand in hand. The subject of armed forces and society must be explored by elaborating the specific structure, composition, and resources of individual military forces within the political context of involvement in crises and within their social environment.

NOTES

1. The Social Science Research Council formed a committee on civilian-military relations, which prepared *Civilian-Military Relations: An Annotated Bibliography, 1940–52.* See also William T. R. Fox, "Civilian, Soldier, and American Military Policy," *World Politics* VII (April 1955), 402–418; and Fox, "Civilian-Military Relations Research: The Social Science Research Committee and Its Research Survey," *World Politics* VI (January 1954), 278–288; Roger Hilsman, "Research on Military Affairs," *World Politics* VII (April 1955), 490–503.
2. For a discussion of militarism see Alfred Vagts, *A History of Militarism,* (New York: The Free Press, 1967); Wilson C. McWilliams, *Garrisons and Governments,* (San Francisco: Chandler Publishing Co., 1967), pp. 1–18; Karl von Clausewitz, *War, Politics and Power,* (Chicago: Regnery Co., 1962); and Friedrich Engels, "The Origin of the Family, Private Property and the State," in Marx and Engels, *Collected Works,* (Moscow: Foreign Languages Publishing House, 1955), pp. 170–326, (Vol. II). A few studies of pre-twentieth-century European militaries are: William O. Shanahan, *Prussian Military Reform, 1786–1813,* (New York: Columbia University Press, 1945); Urbain Gohier, *L'Armée contre la Nation,* (Paris: Editions de la Revue Blanche, 1899); Richard D. Challener, *The French Theory of the Nation in Arms,* (New York: Columbia University Press, 1955), and David B. Ralston, ed., *Soldiers and States: Civil-Military Relations in Modern Europe,* (Boston: Heath, 1966), which deals with both pre-twentieth-century and contemporary periods.
3. Compare the works on militarism in developing areas to Morris Janowitz, *The Professional Soldier,* (New York: The Free Press, 1960); Samuel Huntington, *The Soldier and the State: Theory and Politics of Civil-Military Relations,* (New York: Vintage Books, 1957); Huntington, ed., *Changing Patterns of Military Politics,* (New York: The Free Press, 1962); Samuel Stouffer, et al., *The American Soldier,* (New York: Wiley, 1965).
4. Among the works which have taken a broad look at militaries in new nations and developing areas are: William Gutteridge, *Armed Forces in the New States,* (London: Oxford University Press, 1962); and *Military Institutions and Power in the New States,* (New York: Praeger, 1965); Morris Janowitz, *The Military in the Political Development of New Nations,* (Chicago: University of Chicago Press, 1964); Edward Shils, "The Military in the Political Development of the New States," in John J. Johnson, ed., *The Role of the Military in Underdeveloped Countries,* (Princeton, N.J.: Princeton University Press, 1962); Lucian Pye, "Armies in the Process of Political Modernization," *ibid.;* Samuel Finer, *The Man on Horseback,* (New

York: Praeger, 1962); Huntington, *Political Order in Changing Societies,* (New Haven: Yale University Press, 1968); especially ch. 4, "Praetorianism and Political Decay," pp. 192–263; and Hans Daalder, *The Role of the Military in the Emerging Countries,* (The Hague: Mouton, 1962).

5. Among the country studies of the military are: P. J. Vatikiotis, *The Egyptian Army in Politics,* (Bloomington, Ind.: University of Indiana Press, 1961), and *Politics and the Military in Jordan,* (New York: Praeger, 1967); Gordon Torry, *Syrian Politics and the Military, 1945–48,* (Columbus, Ohio: Ohio State University Press, 1969); Amos Perlmutter, *Military and Politics in Israel,* (New York, Praeger, 1969); Morroe Berger, *Military Elite and Social Change; Egypt Since Napoleon,* (Princeton, N.J.: Princeton University Press, 1960); Robert A. Potash, *The Army and Politics in Argentina,* 1928–1945, (Stanford: Stanford University Press, 1969); and Ellis Joffe, *Party and Army: Professionalism and Political Control in the Chinese Officer Corps, 1949–1964,* Harvard East Asian Monographs, No. 19 (Cambridge, Mass.: Harvard University Press, 1967).
6. For example: Sydney Fischer, ed., *The Military in the Middle East,* (Columbus, Ohio: Ohio State University Press, 1963); John J. Johnson, *The Military and Society in Latin America,* (Stanford: Stanford University Press, 1964); Edwin Lieuwen, *Arms and Politics in Latin America,* (New York: Praeger, 1961); and his *Generals versus Presidents: The New Military in Latin America,* (New York: Praeger, 1964); J. M. Lee, *African Armies and Civil Order,* (London: Chatto and Windus, 1969); and J. C. Hurewitz, *Middle East Politics: The Military Dimension,* (New York: Praeger, 1969).
7. Wilson C. McWilliams, ed., *Garrisons and Governments: Politics and the Military in New States,* (San Francisco: Chandler Publishing Co., 1967); Johnson, *The Role of the Military in Underdeveloped Countries, op. cit.;* Henry Bienen, ed., *The Military Intervenes: Case Studies in Political Change,* (New York: Russell Sage Foundation, 1968); Jacques van Doorn, ed., *Armed Forces and Society: Sociological Essays,* (The Hague: Mouton, 1968); and *Military Professions and Military Regimes,* (The Hague: Mouton, 1970).
8. Bienen, *op. cit.*
9. Samuel P. Huntington, "Patterns of Violence in World Politics," in *Changing Patterns of Military Politics,* p. 36.
10. *Ibid.*
11. *Ibid.*
12. See Bienen, *op. cit.,* pp. 2–4, 100–102, 124–126.
13. Guy Pauker, "Southeast Asia As a Problem Area in the Next Decade," *World Politics* XI (April 1959), p. 343.
14. This point is made by Edward Bernard Glick, *Peaceful Conflict,* (Harrisburg, Pa.: Stackpole, 1967).
15. Marion J. Levy, Jr., *Modernization and the Structure of Societies,* (Princeton, N.J.: Princeton University Press, 1966), Vol. II, p. 574.
16. As Lucian Pye has argued in "Armies in the Process of Political Modernization," in Johnson, *The Role of the Military in Underdeveloped Countries,* p. 70.

17. *Ibid.*, pp. 69–70.
18. Pauker, *op. cit.*, pp. 339–345.
19. In particular see the Annexes to the Draper Report published in Volume II of the *Report of the President's Committee to Study the United States Military Assistance Program,* (Washington, D.C., 1959).
20. See, for example, United States Senate, *International Development and Security,* Hearings, Committee on Foreign Relations, 87th Congress, First Session, Part II, May, 1961, pp. 610–612.
21. For an examination of views about the Latin American militaries see L. N. McAlister, "Recent Research and Writings on the Role of the Military in Latin America," *Latin American Research Review,* Vol. II, No. 1 (Fall 1966), pp. 5–37; these points are from McAlister. Lieuwen, for one, continued to see the military as a conservative organization although he noted changes taking place within the Latin American militaries. See *Generals versus Presidents, op. cit.*
22. Janowitz, *op. cit.*, pp. v–vi.
23. José Nun, "The Middle-Class Military Coup," in Claudio Veliz, ed., *The Politics of Conformity,* (Oxford: Oxford University Press, 1967).
24. Johnson, in *The Military and Society, op. cit.*, put this position forward.
25. See Lyle N. McAlister, "The Military," in John J. Johnson, ed., *Continuity and Change in Latin America* (Stanford: Stanford University Press, 1964), especially pp. 158–160. McAlister does not argue either position himself; he reviews contradictory propositions about the military in Latin America.
26. Levy, *op. cit.*, Vol. II, p. 603.
27. Levy, *op. cit.*, p. 35.
28. Pye, *op. cit.*
29. Morris Janowitz, "Organizing Multiple Goals: War Making and Arms Control," in Morris Janowitz, ed., *The New Military: Changing Patterns of Organization,* (New York: Norton, 1964), p. 29.
30. Janowitz, *The Military in the Political Development of New Nations,* pp. 104–105.
31. Halpern, *op. cit.*, p. 304.
32. Also see, Dankwart A. Rustow, "The Army and the Founding of the Turkish Republic," *World Politics* XI (July 1959), pp. 513–552.
33. Manfred Halpern has argued in a similar way, "The Arab armies failed in the arts of death because Arab society has not yet succeeded in the arts of modern life," in a review of Joel Carmichael's, "The Shaping of the Arabs," *Washington Post, Book World,* September 17, 1967, p. 20.
34. Bienen, *op. cit.*, p. xix.
35. Halpern, "The New Middle Class," *op. cit.*, p. 301.
36. Fred Riggs, review of Marion J. Levy Jr.'s "Modernization and the Structure of Societies," *American Political Science Review* LXI (June 1967), pp. 490–491.
37. For an illustration of disunity in military organizations see Bienen, *The Military Intervenes,* especially Philip B. Springer's "Disunity and Disorder: Factional Politics in the Argentine Military," Jae

Souk Sohn, "Political Dominance and Political Failure: The Role of the Military in the Republic of Korea." For a general discussion of organizational conservatism in armed forces see McWilliams, *op. cit.*, p. 20. As McWilliams points out, organizational conservatism in armed forces has been discussed by W. Kaempffert in "War and Technology," *American Journal of Sociology,* 46 (1941), pp. 431–434; and John U. Nef, *War and Human Progress,* (Cambridge, 1950).

38. Bienen, *op. cit.*, p. xv.
39. McWilliams, *op. cit.*, p. 25.
40. Halpern, *op. cit.*, p. 305. He takes to task P. J. Vatikiotis' *The Egyptian Army in Politics: Pattern for New Nations,* (Bloomington, Ind.: University of Indiana Press, 1961) for neglecting these distinctions.
41. As McWilliams asserts, *op. cit.*, p. 19.
42. Halpern, *Book World,* p. 3.
43. As McWilliams points out can be the case, *op. cit.*, p. 20 and citing Morris Janowitz, "Changing Patterns of Organizational Authority: The Military Establishment," *Administrative Science Quarterly,* Vol. III (1959), pp. 474–493.
44. Aristide Zolberg, "Military Intervention in the New States of Tropical Africa: Elements of Comparative Analysis," in Bienen, *op. cit.* Zolberg utilizes Talcott Parson's distinction between power and force. See Parson's "Some Reflections on the Place of Force in Social Process," in Harry Eckstein, ed., *Internal War,* (New York: Free Press, 1964), p. 59.
45. Bienen, *op. cit.*
46. David Apter has posited this relationship in *The Politics of Modernization,* (Chicago: University of Chicago Press, 1965), p. 40.
47. Edward Shils, "The Military in the Political Development of the New States," in Johnson, *op. cit.*, p. 9.
48. Pauker, *op. cit.*
49. Levy, *op. cit.*
50. Manfred Halpern, *The Politics of Social Change in the Middle East and North Africa, op. cit.*, pp. 283, 285. See also Henry Bienen, "Tanganyika and the Soviet Union: The Party and the No Party State," *Transition,* Vol. III (March-April 1964), pp. 25–32.
51. McWilliams, *op. cit.*, pp. 29–39. See also Gino Germani and Kalman Silver, "Politics, Social Structure and Military Intervention in Latin America," in McWilliams, *op. cit.*, pp. 227–248.
52. Huntington's analysis is set forth in the article we reproduced in this book and also in "Political Modernization: America versus Europe," *World Politics* XVIII (April 1966), pp. 378–414, and in his *Political Order in Changing Societies,* (New Haven: Yale University Press, 1968).
53. Elsewhere, Huntington differentiates between military groups. See "Patterns of Violence in World Politics," *op. cit.*, pp. 32–40.
54. See Sohn, *op. cit.*
55. Nur Yalman in "Intervention and Extrication: The Officer Corps in the Turkish Crises," in Bienen, *The Military Intervenes,* pp. 127–144.
56. *Information and Guidance on Military Assistance* (10th ed.), p. 12, prepared by the Evaluation Division of Military Assistance, Deputy Chief of Staff, S. and L. Headquarters (Washington 1966).

57. General Curtis Lemay, "Strategic Advantage is Key to All Tasks," Air Force Information Policy Letter: Supplement for Commanders, No. 125, Internal Information Division, SAF, 11, Pentagon, cited by Willard F. Barber and C. Neale Ronning, *Internal Security and Military Power: Counter-Insurgency and Civic Action in Latin America,* (Columbus, Ohio: Ohio State University Press, 1962), p. 162.
58. Glick, *op. cit.,* p. 186.
59. C. Windle and T. Vallance, "Optimizing Military Assistance Training," *World Politics* XV (October 1964), pp. 91–107.
60. Glick, *op. cit.,* pp. 185–186.
61. Charles Wolfe, *United States Policy and the Third World,* (Boston: Little, Brown, 1967), pp. 90–109; Robert Putnam, "Toward Explaining Military Intervention in Latin American Politics," *World Politics* XX (October 1967), pp. 83–110.
62. Glick sketches some problems of civic action, *op. cit.,* pp. 175–187; so does Davis B. Bobrow, "Limitations on Civil Action," in Edward Gude, Clark C. Abt, and Davis B. Bobrow, *Arms Control in The Developing Areas,* A Unicorn Study, Phase II Report, for the Directorate for Arms Control, Office of the Assistant Secretary of Defense for International Security Affairs, Contract Number SD–125. Prepared by the Strategic Studies Department, Missile and Space Division, Raytheon Co., Bedford, Mass., pp. 131–141; also Barber and Ronning, *op. cit.,* point to difficulties with civic action programs.

Further Reading

Very important sources of information are the publications of the Institute for Strategic Studies, London.

Bibliographical Essays

Lissak, Moshe. "Selected Literature of Revolutions and Coups d'Etat in the Developing Countries." In Morris Janowitz, ed., *The New Military*. New York: Norton, 1964. Pp. 339–362.

McCalister, L. N. "Recent Research and Writings on the Role of the Military in Latin America." *Latin American Research Review* II (Fall 1966), 5056.

Riddleberger, B. "Military Roles in Developing Countries: An Inventory of Past Research and Analysis." Washington, D.C.: Special Operations Research Office, 1965.

Rouquié, Alain. "Le Role politique des forces armées en Amérique Latine." *Revue Française de Science Politique* XIX (August 1969), 862–893.

Journals

The area studies journals often contain articles on the military. These include *The Journal of Modern African Studies; Africa Report; Pacific Affairs; China Quarterly; The Journal of Asian Studies; Middle East Journal; Survey; Journal of Inter-American Studies; Latin American Research Review.*

The major political science journals also have contributions on the military and political development: *World Politics; American Political*

Science Review; Western Political Science Quarterly; Government and Opposition; Political Science; Orbis; Midwest Journal of Political Science.

A number of journals deal specifically with the military. Among them are *Frères d'Armes; Military Review; U.S. Naval Institute Proceedings; Army; Air University Review.*

Books and Reports on the Military and Political Development

GENERAL WORKS:

Andrzejewski (now Andreski), Stanislaw. *Military Organization and Society.* London: Routledge and Kegan Paul, 1954.

Bienen, Henry, ed. *The Military Intervenes: Case Studies in Political Development.* New York: Russell Sage Foundation, 1968.

Chorley, Katherine. *Armies and the Art of Revolution.* London: Faber and Faber, 1943.

Daalder, Hans. *The Role of the Military in the Emerging Countries.* The Hague: Mouton, 1962.

Finer, Samuel. *The Man on Horseback.* London: Pall Mall, 1962.

Glick, Edward Bernard. *Peaceful Conflict.* Harrisburg, Pa.: Stackpole, 1967.

Gutteridge, William. *Armed Forces in New States.* London: Oxford University Press, 1962.

———. *Military Institutions and Power in New States.* London: Pall Mall, 1965.

Hamon, Leo. *Le Rôle extra-militaire de l'armée dans le Tiers Monde.* Paris: Presses Universitaires de France, 1966.

Huntington, Samuel P., ed. *Changing Patterns of Military Politics.* New York: The Free Press, 1962.

———. *Political Order in Changing Societies.* New Haven: Yale University Press, 1968.

———. *The Soldier and the State.* New York: Vintage Books, 1957.

Janowitz, Morris, ed. *The Military in the Political Development of New Nations.* Chicago: University of Chicago Press, 1964.

———. *The New Military: Changing Patterns of Organization.* New York: Russell Sage Foundation, 1964.

Johnson, John J. *The Role of the Military in Underdeveloped Countries.* Princeton, N.J.: Princeton University Press, 1964.

Lang, Kurt. *Sociology of the Military: A Selected and Annotated Bibliography.* Chicago: Inter-University Seminar on Armed Forces and Society, 1969.

Lutwak, Edward. *Coup d'Etat.* New York: Knopf, 1969.

McWilliams, Wilson C., ed. *Garrisons and Governments.* San Francisco: Chandler Publishing Co., 1967.

VanDoorn, Jacques, ed. *Armed Forces and Society.* The Hague: Mouton, 1968.

———. *Military Professions and Military Regimes.* The Hague: Mouton, 1969.

AFRICA

Afrifa, A. A. *The Ghana Coup.* London: Humanities Press, 1966.

Bell, M. J. V. *Army and Nation in Sub-Saharan Africa.* London: Institute of Strategic Studies, 1965.

Chaliand, Gérard. *Lutte armée en Afrique.* Paris, Prospero, 1967.

Grundy, Kenneth. *Conflicting Images of the Military in Africa.* Nairobi: East African Publishing House, 1968.

Lee, J. M. *African Armies and Civil Order.* London: Chatto and Windus, 1969.

Nkrumah, Kwame. *Dark Days in Ghana.* London: Lawrence and Wishart, 1969.

Wood, David. *The Armed Forces of African States.* London: Institute of Strategic Studies, 1965.

Zolberg, Aristide. *Creating Political Order.* Chicago: Rand McNally, 1966.

THE MIDDLE EAST

Berger, Morroe. *Military Elite and Social Change; Egypt Since Napoleon.* Princeton, N.J.: Princeton University Press, 1960.

Fisher, Sydney N., ed. *The Military in the Middle East.* Columbus, Ohio: Ohio State University Press, 1963.

Halpern, Manfred. *The Politics of Social Change in North Africa and the Middle East.* Princeton, N.J.: Princeton University Press, 1964.

Hurewitz, J. C. *Middle East Politics: The Military Dimension.* New York: Praeger, 1969.

Perlmutter, Amos. *Military and Politics in Israel.* New York: Praeger, 1969.

Torrey, Gordon. *Syrian Politics and the Military, 1945–48.* Columbus, Ohio: Ohio State University Press, 1969.

Vatikiotis, P. J. *Politics and the Military in Jordan.* London: Frank Cass, 1967.

———. *The Egyptian Army in Politics.* Bloomington, Ind.: University of Indiana Press, 1961.

Weiker, W. F. *The Turkish Revolution, 1960–61: Aspects of Military Politics.* Washington, D.C.: The Brookings Institution, 1963.

LATIN AMERICA

Alba, Victor. *El Militarismo: Ensayo sobre un fenómeno pólitico-social.* Mexico City: Biblioteca de Ensayos Sociologicos, Instituto de Investigaciones Sociales, Universidad National Autonoma de Mexico, 1959.

Barber, Willard F., and C. Neale Ronning. *Internal Security and Military Power: Counterinsurgency and Civic Action in Latin America.* Columbus, Ohio: Ohio State University Press, 1966.

De Imaz, Jorge Luis. *Los qe mandan: Las fuerzas armadas en Argentina.* Buenos Aires: Eudeba, 1964.

Fluharty, Vernon. *Dance of the Millions: Military Rule and the Social Revolution in Colombia, 1930–1956.* Pittsburgh, Pa.: University of Pittsburgh Press, 1957.

Horowitz, Irving. *Three Worlds of Development: The Theory and Practice of International Stratification.* New York: Oxford University Press, 1966.

Johnson, John J., ed. *The Military and Society in Latin America.* Stanford, Cal.: Stanford University Press, 1964.

Lieuwen, Edwin. *Arms and Politics in Latin America.* New York: Praeger, 1961.

———. *Generals Versus Presidents: Neomilitarism in Latin America.* New York: Praeger, 1964.

Needler, Martin C. *Political Development in Latin America: Instability, Violence, and Evolutionary Change.* New York: Random House, 1968.

North, Lisa. *Civil-Military Relations in Argentina, Chile, and Peru.* Berkeley, Cal.: University of California Press, 1966.

Potash, Robert A. *The Army and Politics in Argentina, 1928–1945.* Stanford, Cal.: Stanford University Press, 1969.

ASIA

Callard, Keith. *Political Forces in Pakistan, 1947–1959.* New York: Macmillan, 1957.

George, Alexander. *The Chinese Communist Army in Action.* New York: Columbia University Press, 1967.

Griffith, Samuel B. *The Chinese People's Liberation Army.* New York: McGraw-Hill, 1967.

Joffe, Ellis. *Party and Army: Professionalism and Political Control in the Chinese Officer Corps, 1949–1964,* Harvard East Asian Monographs, No. 19. Cambridge, Mass.: Harvard University Press, 1967.

Kahin, George. *Nationalism and Revolution in Indonesia.* Ithaca, N.Y.: Cornell University Press, 1952.

Khan, Fazul Muqueem. *The Story of the Pakistan Army.* Kerachi: Oxford University Press, 1963.

Tinker, Hugh. *India and Pakistan: A Political Analysis.* New York: Oxford University Press, 1962.

Wilson, David. *Politics in Thailand.* Ithaca, N. Y.: Cornell University Press, 1962.

1 *Armed Force Organizations*

MARION J. LEVY, JR.

INTRODUCTION

General State of the Subject

Curiously enough, in academic discussions armed force organizations constitute one of the most generally ignored and overlooked areas of great importance despite the role of armed force organizations in budgets, despite the frequent figuring of members of such organizations as central figures in revolutionary situations, despite the decisive role played by operations in terms of such organizations during periods of war, etc. The vast majority of the work on armed force systems has been carried out by historians or persons interested in armed force systems per se (e.g., military experts) rather than by general social scientists. In recent years an increasing number of social scientists, especially in

Reprinted by permission of the author and the publisher from *Modernization and the Structure of Societies: A Setting for International Affairs,* by Marion J. Levy, Jr. (Princeton, N.J.: Princeton University Press, 1966), Chaper 4, "Armed Force Organization," pp. 571-605. Slightly abridged and truculently agreed to by the author.

the fields of sociology and political science, have concerned themselves with armed force systems as such or with restricted interrelations between armed force organizations and other specific matters.[1] The fact remains that general social treatment of societies or nations in which the armed force organizations are obviously of considerable relevance are on the whole strikingly sparing in their mention of those organizations.

As far as the works of American scholars are concerned, there are in the main two bases for this neglect. The first has to do with a peculiarity of United States society. As will be indicated below, armed force organizations have two major foci of significance in any society. One is the role of such organizations as far as external use, either for defense or aggression, is concerned. The other is the use of such forces for the internal maintenance of order. As far as external use is concerned, until participation in World War II, members of United States society relied overwhelmingly on armed forces which could be called into being after an emergency became clear. The extent of armed force organization maintained during periods of peace was a relatively minor consideration as far as the general membership of the society was concerned. For a whole series of reasons too complicated to go into here, armed force organizations were not in general viewed as having any particular role in the internal maintenance of order. Even following World War II, there was a genuine sense in which it was much easier for Americans to envisage the sending of troops to Korea than to Little Rock. In general, Americans tended to denigrate armed force organizations save those which were considered as essentially citizens' armies in times of emergencies. Regularly and continuously maintained armed force organizations were on the whole suspect. In time of peace we Americans generally tended to regard armed force careers as undertaken only if one were unable to do any better.

To this general American tendency to look down on or ignore armed force organizations as a serious focus of consideration one must add the far more general parochialism of academics with regard to members of the armed forces. There are few subjects about which academics have more stereotypical views than about armed forces in general and the individuals who seek professional careers in terms of them in particular. We generally look down on professional military personnel; we regard them as unintelligent, humorless, authoritarian, and above all anti-intellectual. Our average view of what armed force organizations or professional military personnel are like has little to do with the state of the facts. Actually, there are some amusing departures from the stereotype. A reasonable case may be made out for the fact that, particularly with the developments of modernization, the planners and members of armed force organizations are more likely to put stress on education and on various forms of intellectual training than are the members or planners of any other subsystems of their respective governments. After all, in times of peace, armed force organizations are explicitly organizations in terms of which training and planning take place. In any context in which the technologies on which some operations in terms of armed force organizations depend are constantly changing, sooner or later the members of those organizations must increase their interest in new education at least with regard to the new technological frontiers. In the United States case at least, members of the armed force organizations have given more thought to continuous reeducation of the members of their organizations than have the members of any other subsystem of the government. In the public affairs field generally, Americans as a whole and American social scientists in particular have tended overwhelmingly to think of armed force organizations as at best a necessary evil rather than as organizations to be brought into creative attempts to analyze social phenomena in general. Unless specific crises and specific problems involving armed force organizations in the most obvious and direct sense have appeared, it has simply not occurred to most of us that they were likely to have any relevance.[2]

Specific Interests Here

I make no attempt whatsoever to treat armed force organizations at any length in the context of problems of war, since in that context they have received their most thorough and detailed treatment by military experts, historians, etc. The concern here is rather with possible relevances of armed force organizations in societies generally and in the process of development along increasingly modernized lines—especially the development of latecomers to the process of modernization. The potentialities of such organizations as foci of radicalism and of controlled development have seldom been explored.[3] With a degree of unanimity that is fascinating, both the professional military members of most societies as well as the civilians have viewed the armed force organizations as primarily of relevance in situations in which applications of force are of the essence. No doubt, both the military professionals and the civilian experts are quite correct that this will continue to be the major focus of interest as far as armed force organizations are concerned, but it may at least be interesting to raise questions.

General Setting of Armed Force Organizations

Armed Force Organizations as Objects of Explicit Social Engineering

Armed force organizations, whenever found, appear almost inevitably as explicitly planned and carefully differentiated from other organizations. It is not true that organizations are generally explicit examples of social engineering. It is exceedingly doubtful, for example, whether an organization such as the family is generally an example of social engineering.[4] Whatever the case of organizations in general, armed force organizations are usually the result of explicit planning. Armed force organizations may vary enormously with regard to the range of actions appropriately carried out in terms of them, but they are not generally formed on the spur of the moment, and they are usually quite explicitly organized with a view to efficient functioning of their members in pursuit of relatively well understood ends. There are several reasons for this. In the first place, to a high degree armed force organizations are viewed as means by most of the members of the society most of the time. The means may be offensive or defensive or have to do with the general maintenance of internal order, but the organizations are instrumentally viewed. In the second place, these organizations are always specialized on the basis of sex and age, and therefore they are not organizations in terms of which every individual in the society can more or less casually grow up as a member. They are definitely organizations into which individuals are inducted for fairly explicit purposes. It is perfectly true that there are plenty of examples in social history of armed force organizations which are in a sense only called into being when needed, but in such cases what constitutes needs in this sense are generally planned for and quite explicit. This sort of thing was quite prevalent as a characteristic of feudal societies. There might be no activity in terms of the armed force organizations at a given point in time, but when a given lord called into being the armed force organizations that he had the right to invoke, there where tables of organization in the form of the obligations owed him by his vassals. For example, each owed him a given amount of time or a given number of knights, horses, grooms, and so forth. Throughout the history of mankind few things have been as explicitly structured as religious ritual, magical practices, and armed force organizations.

Two General Foci

Operations in terms of armed force organizations are usually conceived to be oriented toward external *and* internal uses. It is the external applications of the efforts of the members of the armed force organization that tend to come immediately to mind when Americans consider these problems. We think of the members of the armed forces as being bent on aggressive exploits or defensive ones. Rightly or wrongly we have convinced ourselves that only defensive applications of armed forces externally are justifiable. As everyone knows, this attitude on the part of Americans in general constitutes one of the major realities as well as one of the major question marks of present-day international affairs. On the one hand there are those individuals who are thoroughly convinced that Americans are so bound by this point of view that, no matter how certain the knowledge may be that members of another society contemplate an attack, we cannot strike first, no matter how defensively such an initial strike might be conceived. There are also those who maintain that one of the real question marks in international stability is whether or not those in position to give orders to the members of our armed force organizations are in fact still bound by such thinking.

From this point of view both the society of Great Britain and that of the United States have been quite unusual. Neither society has been characterized for well over 150 years by any large-scale uses of armed forces for the internal maintenance of order. The closest approach to this on a large scale was our so-called Civil War, but at least from the point of view of the participants on both sides that has come somehow to be considered as involving external uses of armed forces and not as a use of the armed forces for the maintenance of order, regardless of what the views of an objective observer might be. The views of such an observer have not in any case shaped the attitude of Americans toward the legitimate applications of armed forces. For the members of the societies of both the United States and Great Britain it is in a quite genuine sense easier to contemplate the use of the members of armed force organizations abroad than at home.

For the members of most societies, however, the relevance of armed force organizations for internal uses is as much a matter of concern as their relevance for uses abroad. Generally the armed force organizations are clearly recognized as the main organizations in terms of which force is used for the maintenance of internal stability. One might maintain that even Americans and the English recognize this readily, if one discounts what some might almost consider a verbal distinction between internal opposed to external armed force organizations, the external ones being generally recognized as armed forces in the ordinary sense, the internal ones being recognized as a police force. Even for us, operation in terms of various police organizations is certainly taken for granted as the means par excellence for curbing certain types of deviant behavior. The actual differentiation in organizational terms between *internal* and *external* armed forces is by no means rare, but it is probably unusual for the distinction to be carried to such lengths that the general members of the society regard it as especially shocking and unusual for the armed forces generally associated with external application to be called out in support of what is generally regarded as a problem of internal stability. Perhaps the members of highly stable societies generally regard members of the armed force organizations as having little to do internally. This would be a thoroughly expectable corollary of the fact that whenever actual force is exercised as a day-to-day way of securing conformity, the general internal stability of the society concerned is highly problematical.

The distinction between external and internal uses of members of armed force organizations is further important because in the present world situation no society of any considerable scale is going to be characterized by an absence of armed force organizations—regardless of the form of government or leadership or ideology expressed by the general membership. It is almost certainly true that in the present world situation fewer and fewer armed force organizations are of great significance for what might be called aggressive foreign uses. This may be obscured by all sorts of minor-scale applications of armed forces aggressively by the members of societies of some scale, or in some cases by members of societies of considerable scale.

Much more important than the maintenance of armed forces for defensive foreign purposes is likely to be the continued maintenance of such organizations because of their internal significance for the general maintenance of order. People may cease to regard such organizations as armed force organizations in the ordinary sense, but if one is to judge from the state of the world today, even that seems unlikely. Reliance on members of the armed forces either by themselves or by others as the source of stability of government is quite widespread throughout societies enormously varied in other respects. The maintenance of armed force organizations for these purposes requires some of the same equipment necessary for such organizations conceived primarily in terms of operations abroad or in defense against members of other armed force organizations from abroad. The maintenance of such organizations in societies is not and never has been solely explicable in terms of international involvements. In societies of any considerable scale the maintenance of armed force organizations is always intimately involved with the question of general stability—especially in its political aspects.

In the early stages of the development of relatively modernized societies in the form of nations which prior to that time were to a high degree self-sufficient, one of the early effects was to make possible a sharper public differentiation of internal and external affairs at exactly the same point in history at which the developments were making the maintenance of the distinction meaningless. Today, when increasingly there are no domestic affairs of any importance without international implications and vice versa, the symbol of the last resort in international affairs—the employment of members of armed force organizations—may in the future become increasingly significant in the maintenance of the kind of internal solutions to problems of stability that keep the domestic situation of one nation from involving the intervention of members of another as an act of solution.

Isolation of Armed Force Personnel

One of the most fascinating aspects of armed force organizations is the literal degree of isolation of members of these organizations from members of other organizations in the society. When operations take place in terms of armed force organizations, to some extent their members are always isolated as clearly as possible from nonmembers. The significance of the uniform in this connection cannot be questioned.[5] The isolation takes other forms as well. To some extent members of the armed force organizations live in isolation from nonmembers, especially when what they are doing is conceived to be important in terms of the objectives of the leaders and planners of the organization and especially when they are involved in training.

Armed force organizations may be modernized well in advance of the extent of modernization elsewhere in the social contexts in which they are found. As that modernization proceeds, the importance of isolation for training and maintenance increases.[6] The isolation of armed force personnel means, among other things, that the interrelationships of armed force personnel and other members of the society are periodic and delimited. They are subject to enormous variation in these respects. For quite considerable lengths of time under crisis circumstances or even under training circumstances, members of armed force organizations may be kept in nearly complete isolation from other members of the society. Most of the other organizations in societies with comparable isolation of their members are generally characterized as ascetic organizations and usually as some form of predominantly religiously oriented organization.

The isolation of members of the armed forces is also in terms of the age classification of the members. Even when the members of the armed force organization are determined primarily by birth or class or the like, they generally belong to the armed force organization only for a limited age period. That period is usually all or a portion of their adulthood. This means, of course, that one of the problems universally associated with membership in armed force organizations is the problem of getting members from a general social context and of returning them to a general social context, if they survive to the age of retirement. Within the period of their membership in the armed force organization there may be other periodic returns to a general social context, but in all such cases it is important that the rules governing the transition from one context to another be clear-cut and that the solidarity of the member of the armed force organization with other members of that organization take precedence over his other solidarities in the general social context.

Critical for an understanding of the fit of armed force organizations into general social structure is the fact that their members are frequently isolated for important periods of their time even when other members of the society except for religious ascetics and imprisoned criminals are never so isolated. Moreover, membership in armed force organizations is not ordinarily confused with either religious asceticism or imprisonment except facetiously. Most armed force organizations are notoriously secular. The levels of regimentation in isolation of members of armed force organizations may approach or even surpass those characteristic of a prison, but regardless of personal attitudes toward armed force organizations, they are certainly radically different in kind from systems of imprisonment as such.

Economic Aspects

Almost inevitably armed force organizations are relatively lacking in self-sufficiency. This is true even of those armed force organizations whose members live off the land. The land referred to is usually inhabited by someone else, and a major strategy of modern warfare in some areas of the world is to remove the population ordinarily present in areas in which members of enemy armed force organizations live off the land. The interrelationships of members of armed force organizations with members of other organizations in the society or in other societies is therefore a matter of the greatest strategic significance from the point of view of operations in terms of those systems. Along with isolation and other special characteristics, logistics is a universal military problem.

The logistical problems of large-scale armed force organizations always constitute critical elements of economic allocation in terms of the societies in which they are found. Generally speaking, as far as members of the armed force organizations are concerned, the availability of the goods and services necessary for their efficient operation is someone else's responsibility. Their responsibility is to make use of these goods and services in particular ways. These two responsibilities do not coincide unless the leaders of the armed force organizations are also the leaders of the government of the society. This means that certain problems of economic allocation which are strategic from the point of view of other members of the society are not necessarily matters of concern from the point of view of members of the armed force organizations.

This difference of point of view may give rise to many problems. Members of armed force organizations are supposed to be rational in their uses of goods and services, but the criteria for rationality need not coincide with those which are generally accepted in other spheres of the society. For example, to use the simplest sort of misunderstanding, ordinarily a public official of the United States government who purchases shoes from one firm at a price double that at which they are offered by members of another firm is not only behaving irrationally from the economic point of view, but the irrationality is so extreme that one would ordinarily suspect that an explanation of it lies in graft and corruption rather than in stupidity. On the other hand, an individual member of an armed force organization responsible for supplies may be quite rationally more preoccupied with the dispersion of his sources of supply in a pattern more or less invulnerable to foreign attack, despite radical differences in the cost curves of different supplying firms, than in buying at the lowest possible price. Particularly in the context of relatively modernized societies with their very high levels of interdependency, the different criteria of rationality with regard to the allocation of goods and services of members of the armed force organizations and of others concerned with the allocation of goods and services in terms of the society may become increasingly relevant to an understanding of such very diverse questions as capital formation in general, income levels, etc. These questions are ordinarily not considered to be a major sphere of concern of members of the armed force organizations at all. This in turn poses delicate problems with regard to training members for high office in such organizations. Whether one should attempt to train armed force leaders to be sophisticated economists is by no means a simple question. Its answer is in no sense directly given by the inescapably important involvement of members of armed force organizations in the allocation of goods and services of the society in general.

Political Aspects

Armed force organizations are always predominantly politically oriented organizations, whether regarded from the point of view of internal or external applications of force.[7] Armed force organizations, as such, are usually regarded as instrumental by the members of the society. Although this matter has been relatively little discussed, the development of specialized armed force organizations is certainly one of the earliest examples in social history of an organization specialized along one aspect of social phenomena. Much more importantly, armed force organizations constitute one of the earliest developments and one of the most general developments of organizations in terms of which a considerable part of the action is expected to be overwhelmingly rational in its orientation. Armed force organizations vary widely in the degree of emphasis by members on traditional ways. One of the academic stereotypes of the nature of armed force officials is that they are overwhelmingly conditioned to thinking in nothing but traditional terms. Many instances of technological conservatism and of conservatism, not only in general strategy but also in what might almost be termed the philosophy of strategy as exemplified in such phrases as the Maginot Line attitude, give some basis for this,[8] but the general inferences drawn from such data constitute a vast and misleading oversimplification.

The technology associated with applications of force and questions of tactics and strategy are always matters about which quite important elements of rational emphasis accrue. There are in the experiences associated with armed force organizations many of the elements that we associate so importantly with the development of science, although most academics prefer to think that the scientist and the army man are, intellectually at least, polar types of animals. Nevertheless, associated with the application of technology and the use of force generally in the terms of armed force organizations is consideration of the data, concern with generalizations, comparative analysis, etc. One would expect the emphasis on rationality to be marked under such circumstances. The objectives of leaders and planners of armed force organizations are in general relatively easy to specify, highly material, and in major respects easy to verify. They may be as easily defined as the death or capture of particular sets of individuals, control over the inhabitants of a given area, defeat of enemy army personnel, etc. Military history is full of bonehead plays, but the demonstration of the superiority of the long bow over mounted knights, for example, is the kind of demonstration that is decisive, clear-cut, and readily understood by the most conservative of witnesses. Furthermore, the fate in social history of those who cling to the less rational emphases is also clear-cut. Many have gone into battles determined that they would either kill like gentlemen or not at all, but the vulnerability of these to enemies more preoccupied with killing and less with aesthetics and ethics has been demonstrated again and again. Whatever the moral implications of such demonstrations may be, the implications of them for an emphasis on rationality or its approximation in the conduct of members of armed force organizations while seeking their major objectives have not generally been lost in history or seen in social science.

Interestingly enough, despite the tremendous emphasis throughout history on armed force organizations as the last resort of government leaders and the like, in the general structures of governments armed force organizations are rarely more than instruments in readiness for emergencies. Rarely do they become the direct focus of government. It has become popular for some to prattle about the power elite and the role of leaders in armed force organizations in that. In various highly authoritarian societies such as that of Germany under the Nazis, Russia under its present rulers, and the like, the armed force organizations have certainly been important centers of attention. Nevertheless, in none of these cases have the armed force organizations as such constituted the major focus of governmental leadership. In most societies armed force organizations have been instruments of governance rather than the government. Most of the cases in which the armed force organizations have been the general focus of the government have been relatively short-lived. Some cases have been spectacular. The leaders of the military organization took over control of the Japanese government in the late 1930s. In the period of Tokugawa Japan (1603–1868) that society had a government whose leaders explicitly consisted only of militarists, but actually this constituted one of the major jests of history, since they were militarists *(bushi, samurai)* who were explicitly forbidden to engage in military endeavors in general and who in fact constituted the civil bureaucrats of their various feudal leaders. In general the men and matériel associated in terms of armed force organizations have considerable value in taking over the control of the government of a society and in making the arrangements for another type of government and/or another set of leaders of the government. Professional soldiers and their subordinates have not been conspicuously successful in social history in long-run governmental administration conducted primarily in terms of armed force organizations.

This is exactly what one would expect if one considers the general emphasis characteristic of such organizations on rationality in the exercise of force. It is the essence of understanding the general political structure of a society as a whole that the allocation of power and responsibility cannot consist overwhelmingly of applications of force and threats of force. Applications of force and high levels of coercion constitute extremes in one direction. Preoccupation about the rational application of such extremes tends to preclude consideration of other measures. In the context of all viable armed force systems there is an emphasis on *command* which is never easily identifiable with an emphasis on *governing* in general.

The frequent emergence of armed force organizations and their leaders in central governmental positions in those societies whose members are late-comers to modernization is not a contradiction to the above. It is rather a function of the fact that, because of the general interest in armed force organizations and the importance of their being modern, the armed force organizations of such societies are likely to be foci of modernization and ladders of social mobility for talent, even if other organizations in the society are not.[9]

Armed Force Organizations as Such

The Hierarchical Nature of Armed Force Organizations

There is a sense in which armed force organizations constitute perhaps the purest case of predominantly politically oriented organizations. Applications of the extreme form of power always constitute the main focus of action in terms of armed force organizations, ideally speaking at least. If these organizations are effective as a set of structures in terms of which individuals pursue such ends, the structures of the organization must stress allocations of power from the top down and allocations of responsibility from the bottom up. It is of the essence of armed force organizations that they are strictly speaking hierarchical in the purest sense of the word. It has already been pointed out that, regardless of ideologies about egalitarianism, most organizations are in fact hierarchical. The hierarchical nature of other organizations may or may not be stressed, may be carried much further actually than ideally, may not be fully appreciated by the individuals involved, etc. Most of the hierarchical characteristics of armed force systems are explicit, ideally speaking, and they tend actually to be clear-cut too. There are some departures from this emphasis on hierarchy. Leaders of some special units, for example, make much of the fact that relatively little attention is paid to rank among their members, but on the whole it is considered a mark of derogation of any armed force organization to imply that the chain of command is unclear. Emphasis on hierarchy is frequently associated with emphasis on tradition, authoritarianism, etc. These are frequently characteristic of armed force organizations too. Nevertheless, it is of the essence of understanding the importance of hierarchical structures in armed force organizations to realize that, given the goals of members and nonmembers alike, a concern for rationality would require an emphasis on hierarchical organization, even in the absence of other justifications of "rank" by the members of armed force organizations.

Rationality

The emphasis on rationality may be especially high relative to that characteristic of the rest of the society. To some extent there is always an explicit emphasis on rationality in the action carried out in terms of armed force organization, no matter how traditionally oriented general relationships may be in general in a society. At least one reason for the emphasis on rationality is that the members of such organizations are always concerned with instrumental uses of weapons and physical skills in forms easily understood rationally. Efficiency in applications of force via tools (i.e., in this case, weapons) is most immediately and easily recognizable.

This heavy emphasis on rationality carries with it an actual emphasis of predominantly universalistic criteria, despite the fact that, ideally speaking, armed force recruitment may be overwhelmingly particularistic. For example, only individuals of a given class by birth are eligible to be military personnel in some cases. Nevertheless, once the organizations are formed, some relatively clear-cut delineations among the members on the basis of their actual or alleged military achievements are inevitably made.

Probably no armed force organization has ever existed without being featured by contests of skill, tournaments, etc., among its members. Rank in terms of authority is often not clearly dependent upon rank in terms of these skills, but they are never irrelevant to one another either. The emphasis on predominantly universalistic criteria, however, may be largely a function of the actual structures rather than of the ideal structures in many cases.

Bureaucratic Aspects

Armed force organizations, regardless of the general social setting, are likely to be pronouncedly bureaucratic in character. Even when they are not highly bureaucratic, the emphasis on bureaucratic qualities of the armed force organizations is likely to be much greater than elsewhere in societies not characterized by the general development of bureaucratic organizations. Whenever the development of bureaucratic organizations is general, such armed force organizations as exist are highly bureaucratic. These organizations to some extent are always characterized by specialization of functions, by great emphasis on office and on the hierarchy of authority and responsibility, by being regarded as highly instrumental, etc.

The bureaucratic character of armed force organizations is especially important in considering the problems of modernization of relatively nonmodernized societies. One of the inevitable changes in any society with an increase in the emphasis on modernization is an increased proliferation of bureaucratic organizations. In some relatively nonmodernized societies the armed force organization may be the major precedent for a bureaucratic or semibureaucratic experience.

The Problem of Control of Armed Force Personnel

The problem of control of the general membership of armed force organizations is usually minimal as such problems go in societies. This is a function of the fact that the general socialization of members of societies is such that membership in an armed force organization is always viewed as something of a thing apart and in some sense in the service of the general membership rather than as an end in itself even in the most militaristically oriented of societies. Despite the fact that new members of armed force organizations receive additional socialization in terms of the organization itself, since they do not ordinarily join prior to an advanced level of maturity, their general socialization is highly relevant. Without general habituation of the members to hierarchial structures of authority and responsibility, it is difficult to see how discipline after joining could account completely for the relatively minimal problem of control of the general membership. Even in the United States, for example—and the general members of United States society have on the whole been about as actively anti-military-organizations as the members of any society in world history—there has been little problem of control of the general membership of armed force organizations.[10] Despite all the griping characteristic of American soldiers, we have lost more members of our armed force organizations by psychological discharges than by mutiny.

The vast majority of problems of control over members of armed force organizations are problems of palace revolutions rather than of mass mutiny. Mass mutinies are exceedingly infrequent. The famous Russian Prince Potemkin incident may have been such a case, but they have rarely occurred save under conditions of general social breakdown proceeding far beyond the armed force organizations themselves. Indeed, one may generalize as follows: The problem of control of the general membership of an armed force organization never reaches considerable proportions save under conditions of extremely general breakdown of the society of which the armed force organization is a subsystem.

The seizure of general control by one quite restricted set of the members of an armed force organization—almost inevitably its leaders of the moment—is a much more general problem in social history. The problem of military-led palace revolutions as a problem of control of armed force personnel is likely to be a problem when there is general social breakdown. It is also a problem under relatively stable conditions if recruitment or general operation in terms of armed force organizations has somehow managed to concentrate opportunistic personalities. Because of the extreme efficiency of applications of force for limited purposes, illegitimate acquisition of control of the armed force organizations always constitutes a considerable seizure of power in and of itself, but it always also constitutes a potential avenue to power in terms of the general society. After all, if a given individual seizes control of the armed force organization in an illegitimate fashion, it is necessary for him to alter some of the general structures of government even if he is merely to maintain the position he has illegitimately seized within the military organization. Thus palace revolutions within military organizations are quite likely to turn into palace revolutions for the society as a whole or in some cases into governmental revolutions as well. In extreme cases they may even turn into general social revolutions.

Far more serious as problems of general control are those of combating states of apathy or the general maintenance of morale. Military specialists for centuries have been well aware of the fact that no equipment and no numbers give the members of any armed force organization superiority over others if they lack the will to fight. It is the fate of the wisdom of the ages to turn into the clichés of the present, if that wisdom is not continually supplemented. One of the great hopes of the development of science was that these developments presaged continuous additions to the wisdom of the ages. In the military field at least the social sciences have not borne out this promise. The major problems of control in the context of armed force organizations are closely related to problems of social control in terms of the society as a whole. Mass apathy of armed force personnel, like mass mutinies, accompanies general social disorganization associated with the society in which these armed forces occur. Given general social disorganization, leaders of large-scale organizations of armed forces cannot overcome problems among their own members, unless the level of isolation of their members from the general membership of the society can be radically increased and rigidly maintained. And even this is not likely to be successful. Indeed, one of the most important forms of dependence or interdependence of armed force organizations and the rest of the society is their extreme dependence on the general social organization for certain basic levels of morale in terms of which members of their own organization can operate.[11]

Insofar as problems of control of members of armed force organizations are not the result of general failures in socialization of the members of these organizations but are rather of the type of restricted mutinous outbreaks, the deviant behavior is extremely vulnerable to the use of secret police techniques. Such activities are likely to contain a high element of rational calculations, and that implies that they are to a high degree subject to rational prediction and rational offset. If secret police are necessary for the control of such phenomena, however, the set of secret police themselves become a major focus of the possibility of palace revolutions within the armed force organizations. Consequently they hold forth also the possibility of palace revolutions in the general government of which the armed force organizations are parts, and depending on the implications of that sort of palace revolution, their revolutionary implications may spread to the governmental level or even the general social level.

Consideration of the general problem of control of members of armed force organizations calls attention once again to an extremely general osbservation about subsystems of societies. Various organizations within a given society may become highly specialized and highly differentiated from one another. If an organization is of any considerable scale, however, and/or of any considerable stability, the variations possible in the social structures which characterize the organization are a function of the general social system in which they exist as subsystems. The general social structures constitute, as it were, the most crucial element in the general setting of any particular organization. They may seem to be a contradiction in the fact that armed force organizations are overwhelmingly oriented to applications of force and the general membership of the societies in which they are found may be much less specifically concerned with such applications. Nevertheless, armed force organizations are as much a mirror of general social structures as any other subsystem in a given society. Our armed force organizations are no less characteristic of our society than are our universities, our churches, or legislative bodies, our courts, or our families. The detailed organizational components of these systems constitute an important component in what one might consider their optical properties if the mirror metaphor is pursued, but the range of optical properties is never infinite and the properties reflected are never solely functions of the mirrors themselves. For those students of public affairs who feel that it is important that armed force organizations be kept under strict control and guidance, the importance of nonarmed force responsibility for the behavior of the members of armed force organizations cannot be ignored.

The utilization of the members of armed forces in the maintenance of general social internal control raises another problem. The more action in terms of armed force organizations is relied upon for general social control, the greater is the likelihood that the leaders of the armed forces will take over the leadership of the government as a whole. It is quite possible for leaders of the armed forces to take over direction of the general government under other circumstances. The history of Japan in the period of the 1930s is clear enough evidence of the possibility. Nevertheless, insofar as the armed force organization becomes the major organizational bulwark of general social control—not as a last resort but as a daily instrument—the general government is likely to become a subsystem of the armed force organization rather than the reverse. There are fairly obvious reasons for this. The day-to-day invocation of an organization overwhelmingly oriented to applications of force for the maintenance of general order is not likely to be the case unless there is a fairly general situation of social disintegration. The purely sadistic application of force generally cannot be long maintained in terms of any society. Under circumstances of such general and continuous application, it is impossible to maintain balances of power and responsibility, if the leaders of the armed force organization do not take over the general policy decisions. There is nothing inevitable about the balance of power and responsibility, but if it is not maintained, there will be widespread social change, usually of a revolutionary sort. If it is under the circumstances just mentioned it can be maintained only if the members of the armed force organizations take a larger role in the domination of the general governmental system.

Isolation of Armed Force Personnel

In the discussion of the general setting of armed force organizations I pointed out that some isolation of armed force personnel is ubiquitous but not ordinarily associated with either asceticism or imprisonment. The generally accepted kind and level of isolation of armed force personnel is extraordinary and it becomes the more extraordinary the more modernized the society.[12] There are many reasons for the relative isolation of armed force personnel. There is the obvious factor involved in the importance of twenty-four hour readiness, and vital to this is the availability of the members without need to separate them out of other sets of individuals. To some extent these ideas tend to fade into the background if the general situation is not regarded as one of emergency. Even if there is no felt emergency, however, another factor is generally in operation which carries with it the implication of relative isolation. This is the inordinate emphasis on training. The skills emphasized in the context of armed force organizations are generally speaking of a high order, and they degenerate rapidly if not constantly exercised. Not the least of these skills is the skill of coordination of large numbers of individuals under critical conditions. If action in terms of armed force organizations is to be effective, it is necessary not only to isolate new members of the organization from the general civilian populace in order to teach them new skills and to integrate them with the going members of the organization, but to some extent it is also necessary to keep them relatively isolated during their membership in order to keep their skills at the readiness level.

The implications of this isolation are quite far-reaching. Not the least of these implications is the inevitability of cliques based largely on armed force membership as such. If general civilian attitudes isolate the members of the armed force units still further, the relevance of cliques as a special form of particularism in armed force organizations is heightened.[13]

Given the general isolation of armed force personnel and given the special emphasis on this isolation until basic training, at least, can be imposed upon the new members, the general social origin of the personnel has a special relevance for their behavior while members. After all, their origins constitute the source of the nonarmed-force opinions of the members of the armed force organizations. These opinions are of great importance when members of the armed force organizations operate in contexts other than those strictly military ones. Within the context of each of the societies in world history the social origins of the personnel concerned is likely to be relatively uniform. For some societies this uniformity results from the fact that only those with a given class standing could become members of the armed forces, as it were. For other societies, insofar as membership in the armed forces is generally open to members of the society, the vast majority of the members of a given society are likely to be individuals of a single class, and hence if the armed force organizations are large scale, most recruitment is likely to come from people of more or less common origins. This is especially true, of course, of relatively nonmodernized socieities. In such societies armed force organizations are frequently elite organizations whose members are likely to come from representatives of a single elite class. If they are not elite organizations and recruitment is open class, the vast majority of individuals concerned are likely to come from agrarian social backgrounds.

The armed force organizations of relatively modernized societies and the new ones of societies whose members are latecomers are overwhelmingly likely to be based on some structure of "universal conscription." For the late-comers this means that the membership is overwhelmingly likely to consist of individuals with peasant background, while members of the officers corps may be of middle-class background. Even this is on the whole unlikely since it is difficult to recruit a large-scale officers corps from persons of middle-class background because of the extreme bimodal character of the social structures of those societies prior to the beginnings of modernization. In an increasing number of cases, whatever the origins of the officer group may be at the outset, in the long run the officers corps is likely to consist increasingly of persons originally brought into the organization by general conscription and elevated to the officers corps on the basis of predominantly universalistic criteria. Thus, increasingly in these social contexts both officers and enlisted personnel are likely to have quite common views of what the general social structure and social problems of their societies are. During the years of their most intensive training in terms of the new organization, they are likely to be, relatively speaking, out of contact with the general social settings to which they were previously accustomed. Therefore, if they get into discussions of general social problems outside strictly military ones, they are likely to have a uniform point of view that, though it may not in fact be a function of the military discipline or of any explicit indoctrination to which they have been subjected, is likely to be so uniform as to be mistaken for a function of regimented experiences in terms of armed force organizations.

Finally, this isolation keeps the members of armed force organizations separated from the general population of the society while the views, discipline, and knowledge of the members change, if and as they do. This is important both for them and for the general membership of the society from which they come. Since the members of armed forces are always to some extent kept in isolation, it is possible to minimize or control the feedback of the introduction of the relatively modernized structures in terms of which they act to the general population at large. In the context of both modernized and relatively nonmodernized societies in process of modernization there is no question but that the new training will be on as highly modernized a basis as possible. I have tried to point out again and again that there are always reasons why elements of modernization are never completely inexplicable to members of even the most nonmodernized societies. In this chapter I have already tried to indicate that, regardless of how traditional the orientations of the members of a given society may be, the emphasis on rationality by members of armed force organizations is likely to be carried to great lengths —that efficiency in weaponry, for example, is not difficult for people to understand even though they may not even have conceived of a given technology of weaponry before. In decisions on the type of training and the type of technology to be put at the disposal of members of the armed force organizations in such societies there will be few arguments about the degrees of modernization—as high a degree as possible will be aimed for. What is unusual in this situation is the extent to which the spread or impact of structures of modernization beyond the focus of immediate introductions can be restricted and controlled if the situation is clearly understood. The implications of this factor for armed force organizations in societies which are being modernized is sufficiently great to justify the devotion of the remainder of this chapter to this question.

Armed Force Organizations in Societies Undergoing Modernization

The Universal Presence of Armed Force Organizations in Such Societies

I asserted at the start of this chapter that the analysis of the armed force organizations has been systematically affected by the American attitude that armed force organizations are necessary only in times of external crisis in particular, and the almost universal academic biases with regard to armed force organizations and their members in general. In the following discussion I do not wish to get involved in questions of whether militarism is good, bad, or indifferent, but I do wish to point to some aspects of the special relationships between armed force organizations and societies in the transition to relatively modernized states.

Armed force organizations will be universally present in societies undergoing modernization regardless of whether or not their leaders and members take into consideration the radically diminished degrees of freedom open to them today for the employment of armed force organizations as instruments in external relations. They will in some form or other be universally present if only because of the importance of such organizations as instruments of internal control during a period of social change in which problems of internal control are likely to be maximized, regardless of how quiescent or turbulent the traditional basis from which change takes place may have been. It is extremely unlikely that armed force organizations be maintained only for this purpose. The external role of armed force organizations is likely to be a matter of prominent consideration, even if the leaders and members of these societies only contemplate defensive applications externally.[14]

Maximal Modernization

No one contemplating an armed force organization wishes less than the most modern weapons. It is sometimes possible, though difficult, to convince the late-comers that the efficiency of modernization may actually be radically increased if initially they do not seek the most modern form of machines because of their comparative advantages for purposes of international trade in other combinations of resources. It is not possible, nor is it generally attempted, to convince them that less than the most modern weapons will serve them best in the planning of the development of their new and modern armed force organizations. When

I hear those who deal with problems of modernization in relatively nonmodernized societies complain about the difficulty of getting certain ideas across with regard to modern technology, I wonder why we have not capitalized on the much greater ease of getting across these ideas in the military sphere. Because of the ease of understanding of certain of the aspects of modernization insofar as it has to do with military transport and weaponry and because of the generally great emphasis on rationality in this sphere regardless of what the previous society has been like, the emphasis on modernization in general with regard to armed force organizations is an emphasis on modernization which generates little resistance. When one speaks of modernizing a government or a city, or in some cases even of agricultural production, there are always fundamentalist objections. There are even fundamentalist objections when one speaks of this with regard to the armed forces, but these objections are relatively rarely persistent or difficult to cope with.[15]

With weaponry as with anything else, modernization cannot be imported piecemeal at the whim of the modernizers. The easy recognition of the importance of modernization in weaponry and logistics in general means, however, that there will in general be an emphasis on the maximum amount of modernization as fast as possible in the development of the new armed force. In this respect the emphasis on modernization may be even greater in the case of some of the late-comers to modernization than in some of the more highly modernized societies. In the former, vested interest in the older technologies are so hopelessly out of date as not to pose a serious barrier to modernization. In the latter the gap between a somewhat older technology and more modern ones is sometimes not sufficiently obvious to make vested interest in older ways appear patently ridiculous.

Universal Conscription Structures

The new basis of recruitment for the armed forces of all societies undergoing modernization will probably be universal conscription. The recruitment procedures may be characterized by graft and corruption, though if universal conscription is attempted this is less likely. Acceptance of personnel will, ideally speaking, be based on the predominantly universalistic criteria of physical fitness and mental ability. The general population from which conscripts will be drawn will be too poverty stricken to accumulate any considerable amount of economic resources to be used for graft and corruption. Furthermore, the increased survival of children accompanying new medical technology will remove some of the earlier resistance to conscription of any sort.[16]

Given the relevance of rationality about physical fitness and mental ability in the context of armed force organizations, it is not difficult to achieve relatively high levels of predominantly universalistic criteria based overwhelmingly on physical fitness and mental ability. Thus, the armed force organizations are likely to have as their new membership not only a cross-section of the general population but also a cross-section of the best talent available from such sources. It is part of the easy rationality of military matters in general that neither the sick nor the stupid make the best recruits. In the context of relatively modernized armed force organizations the emphasis on health is certainly not diminished, and the emphasis on mental ability is if anything increased radically. In addition, the easily comprehended relevance of these factors is as much as anything else likely to dictate a structure of universal conscription rather than some more particularistically based form of recruitment. The relevance of predominantly universalistic criteria for armed force organizations is no more difficult to understand than is the relevance of rational modes of thought.

High Social Mobility

Armed force organizations in societies undergoing modernization will become major ladders—perhaps *the* major ladder—of social mobility. To the extent that these organizations are large in scale they cannot be confined to membership related via kinship to the elite of a relatively nonmodernized society. This is not a question of whether or not the elite care to maintain such a monopoly. There are simply not enough of them to man the armed force organizations let alone their other preferences. If physical fitness and mental ability are emphasized as a basis for original acceptance, they will be emphasized as a basis for promotion in terms of the organization so formed.

Prior to increased modernization, any position of importance in the armed force organization of a relatively nonmodernized society is probably an elite position. Frequently the vast majority of the members of a society are considered too lowly to even aspire to such positions. If they are able to join the armed force organizations at all, ordinarily they join only in positions of extreme subordination, and mental ability or physical fitness do not necessarily promise advancement.[17] Thus, in many contexts entrance into the armed forces alone is taken to be social mobility. In the effort to maximize modernization of armed force organizations, predominantly universalistic criteria for admission will be maintained for advancement.[18] If mobility is not on predominantly universalistic grounds, the price paid in terms of efficient, loyal, and dependable armed force personnel will be heavy.

If the armed force organization is important to the general government for maintaining internal order or for other reasons, high mobility in terms of the armed force organization means high mobility in strategic areas of control as far as the general society is concerned. The case of the modern Japanese Army may be the most extreme case. Nevertheless, in modern history the number of individuals who have come into positions of important leadership in their various societies via roles in the relatively modernized armed force organizations which their countrymen have attempted to set up—never mind what they had in mind in so doing—is far too great to be ignored. If promotion is predominantly universalistic and recruitment is on the basis of universal conscription, the large-scale armed force organizations are more likely than any other single avenue to be ones in terms of which, both absolutely and relatively, the possibility of general social mobility from the most humble levels to the pinnacles of prestige and power is offered and realized.[19]

Armed Force Organizations as Foci of Radicalism

If the use of the term *radical* is not restricted to the clichés of the political right and left, armed force organizations in societies undergoing modernization are likely to be foci of radicalism. The form such radicalism takes is more likely to be explicable in terms of the social backgrounds of the personnel of the armed forces than in terms of the general structures of the armed force organizations. The case of Japan is, again, an excellent illustration. Prior to the development of World War II the major focus of radicalism in Japanese society had nothing to do with communism or socialism in the ordinary sense. The major focus of radicalism in that society was national loyalty—patriotism. The most significantly radical personnel were to be found as members of the armed force organizations, and the leadership resided primarily in young officer cliques. These young officers were precisely examples of high social mobility in terms of the armed force organizations. Almost to a man they came from peasant backgrounds. Insofar as they were dissatisfied with Japanese society as they viewed it aside from their military environment, they were likely to think in terms of the problems which they considered typical of the agrarian peasant environment. Distribution of the fruits of modernization to the peasants may or may not have lagged, rate-wise, behind that to other members of the society, but those individuals constituted the vast bulk of the membership of the society, and such increased benefits as they received were not so dramatic as those in many more restricted urban environments.

The young officers behaved radically as far as the stability of the existing social structure was concerned. To describe them as of rightist opinion is to misunderstand their position thoroughly, for it implies that in general they were conservative and supporters of the business and financial leadership of their society. Verbally, the subjects of their bitterest denunciations and, concretely, the targets of their bullets and the objects of their most spectacular exercises of sword-play were expressly members of the *zaibatsu* (i.e., financiers and businessmen), whom they regarded as betraying the true interests of the Japanese nation, and other military leaders whom they regarded as cravenly subservient to the *zaibatsu*. Their ideology was certainly not Marxist, but they did see the whole nation as betrayed and injured by the comparative disadvantage of the main body of Japanese—the peasants—relative to the always-suspect city folk.

The important thing about the Japanese case is not whether these particular oversimplifications hold true, but rather that as an organization the Japanese Army constituted one of the major foci of radicalism within the society, and in this case at least it constituted *the* focus of radicalism which at the end of the 1930's saw its leadership control Japanese society as a whole. In other societies undergoing modernization, if universal conscription on a predominantly universalistic basis takes place, if promotion of those conscripted continues to be on a predominantly universalistic basis, and if modernization of the armed force organization is maximized, this picture of an armed force organization as a major focus of radicalism will be repeated in every case.

Armed Force Organizations as a Special Focus of Modernization

Preoccupation with armed force organizations as instruments of offense or defense relative to the members of other societies has diverted attention from armed force organizations as the most efficient type of organization for combining maximum rates of modernization with maximum levels of stability and control. The new technology will be taught to the members of the new armed force organizations, and leaders will try to maximize modernization. Furthermore, military technology is not radically different from the technology that is in general characteristic of modernization. Specifically, one may teach the new army recruits, for example, to drive and handle military trucks, but when they have learned their lessons, they have learned to handle trucks in particular, and beyond that they have learned a considerable amount about modern machinery in general. Indeed, if anything, the teaching of technology for military purposes puts greater emphasis on learning it well than does the teaching of technology within a more peaceful context. Within the military context one not only has to learn how to run a truck but also what to do with it in case of a breakdown under critical or near critical circumstances. In ordinary civilian contexts the leeway for less thorough training is much greater even (or, perhaps, especially) in the context of relatively modernized societies.

Not only will modernization in general be taught, however narrow the preoccupation may be with military technology as such, but the teaching will also be done outside the general social context. Here the isolation of the members of the armed force organizations is of the greatest significance. This isolation not only minimizes the impedance of the effect of modernization on the men directly submitted to it. It also minimizes the feed-back to the general social context. Both are of the utmost importance in modernization problems. The former is important because modernization outside such a restricted context may very well raise all sorts of considerations of fundamentalist reactions. Do the people concerned wish to see the new ways imported in this fashion? All manner of soul searching may ensue. Such soul searching is never in the last analysis able to stop the impact of modernization, but it can certainly impede the efficiency with which it is introduced. The isolation and extreme emphasis on authoritarianism characteristic of an armed force organization, the general lack of problems of controlling the individuals who are members of such an organization, and the like—all unite to minimize these problems.

The minimization of the problem of feed-back to the general membership of the society as a whole is perhaps even more important. No matter how isolated the members of the armed force organization may be kept, the spread of the impact of modernization beyond the membership of the armed force organization itself can never be eliminated. Nevertheless, feed-back to the general membership can be considerably restricted, especially if one deals with the members of armed force organizations explicitly with this in mind. The emphasis on modernization in terms of armed force organizations makes it possible to maximize the impact of modernization for the member of an organization in terms of which control over individual behavior is both maximal and simple. The general spread of such structures among the general members of the society, before controls for modern purposes are available and before some of the elements of modernization are better understood than formerly, can be impeded even though it cannot be eliminated. Thus the armed force environment makes possible the maximization of the impact of modernization while minimizing the uncontrolled spread of modernization with the consequent production of problems and crises presently difficult if not impossible to foresee.

In the course of these developments—regardless of whether they are carefully planned—the general membership of the armed force organizations always become a pool of skilled labor in the general modernized sense and not merely with regard to military tools. The officers or leaders of the armed forces always become a civil service of sorts, a set of skilled bureaucratic administrators, whether they like to think of themselves as such or not. Modern military leaders are primarily administrators rather than fighters.[20]

The focus on armed force organizations as things apart from general civilian concerns—as a necessary evil at best—has obscured attention to what may be the most efficient vehicle for the maximization of modernization with a minimization of the uncontrolled spread of side effects. In the past students of public affairs and men of public responsibility have thought in terms of armed force organizations as vehicles of offense and defense or organizations for the maintenance of internal order or some combination of the two. Their impact on the general process of modernization is generally taken to be peripheral at best. This may turn out to be one of the great mistakes of the twentieth century—one of the neglected opportunities of history.

NOTES

1. See for example, Morris Janowitz, *The Professional Soldier* (Glencoe, Ill.: The Free Press, 1960); Samuel P. Huntington, *The Soldier and the State,* (Cambridge, Mass.; Harvard University Press, 1957); John W. Masland and Laurence I. Radway, *Soldiers and Scholars, Military Education and National Policy,* (Princeton, N.J.: Princeton University Press, 1957); and John J. Johnson, ed., *The Role of the Military in Underdeveloped Areas* (Princeton, N.J.: Princeton University Press, 1962).

   Professor Ely Chinoy does refer to the significance of armed forces in his elementary text *(Society, An Introduction to Sociology* [New York: Random House, 1961], pp. 263–265) and gives students initial bibliographical aid on this point.
2. There are, of course, some exceptions. See, for example, *The Role of the Military in Underdeveloped Countries,* John J. Johnson, ed., (Princeton, N.J.: Princeton University Press, 1962), published after this chapter was written.
3. Again, as an exception see *ibid.,* footnote 4.
4. I feel quite sure, in fact, that family organizations in general were not examples of social engineering, although certainly there are many cases in which specific structures of family organizations have been explicitly modified by social engineering. For example, in the period of Ch'in Shih-huang-ti in Chinese history, explicit rules calling for the equal inheritance among all sons were set up in order to combat the continuous maintenance of large accumulations of land and the like. This one bit of explicit social engineering in family terms was quite enough to spell an end to feudalism in Chinese history. This particular bit of social engineering had quite far-reaching implications for Chinese society in general and Chinese family structures in particular.
5. There is not space here to go into any extended analysis of the significance of the uniform in armed force systems, but it is a subject virtually untouched by modern social analysis. In all sorts of ways the uniform is used to epitomize distinctions in office ("Salute the uniform not the man!"). Appearance in or out of uniform in critical situations is prima facie evidence of the most serious differences in intent (a soldier captured in enemy territory in uniform is supposed to receive honorable treatment of a given sort; a soldier of an enemy force captured out of uniform is likely to be shot as a spy). The uniform in armed force systems is a specific instance of a very general social phenomenon of great significance but is little appreciated in public affairs. This is the general phenomenon of making social distinctions literally and easily visible. Dress may not make the man, but it is almost inevitably an enormously helpful item in classifying him.

6. For modern armed force purposes, the old United States conception of minutemen will not work, and other forms of emergency levies won't work either. Even members of the United States society have learned that there must either be a peacetime armed force organization of more than token quality and scope or none at all. Until the beginning of World War II our attitude was generally that in time of peace the more complete the disbandment of members of armed force organizations the better. Something of this attitude was still perceptible in the rapid demobilization of the United States armed forces following the apparent settlement of world problems by the surrenders of Germany and Japan. The views of the place of armed force organizations in American life have changed radically since that time.
7. The political aspect of social phenomena has been defined as having to do with the allocation of power and responsibility, and applications of force have been specifically delineated as an extreme in one direction of these allocations. Armed force organizations are always oriented toward applications of force.
8. The actual role of conservatism and tradition in general in armed force organizations has not, as far as I know, ever been given a general examination by a modern social scientist. Again I would observe that until quite recently only historians and military experts as such have paid a great deal of attention to armed force organizations. Much of the literature has been highly polemic in its orientation. Not only are modern studies by social scientists just beginning and general studies of the relevance of armed force organizations for other social phenomena still virtually nonexistent, but fascinating clichés are accepted almost universally without critical examination. I am not qualified to pass on whether or not the technology available early in World War II was sufficient to crack the Maginot Line, as that line was presumably planned and heralded as invulnerable to the technology of its day. As far as I can make out, however, the line itself was never completed in the form called for by the presumed claims of invulnerability. Its failure in the defense of France was not the result of the kind of frontal assault through or over the line which would have been necessary had the line been completed as conceived. Rather, German military personnel went through areas not covered by the line, "turned it" as it were. If this is the case, the extent to which the experience of early World War II technology forever discredited the Maginot Line psychology is by no means a simple matter to determine. What is more important, the facile acceptance of this as a cliché may actually beg the question of what was wrong with the planning involved. It is one of the grim ironies of history that by the end of that war technological improvements had made the issue, as generally argued, quaint.
9. This matter is pursued in pp. 71–73 of this chapter.
10. The general anti-military-organization sentiments of United States citizens should not be confused with corresponding sentiments against violence in general. The social history of the United States has shown no recalcitrance to violence, but the emphasis has been on individualistic applications of violence rather than carefully regimented ones. Even mob violence places only short-run emphasis on organi-

zation. Within the context of large-scale armed force organizations, Americans are likely to single out as heroes ones who can be understood as individualistic virtuosi of violence such as the famous Sgt. York rather than geniuses of military organization. The latter, with rare exceptions, do not achieve great popular appeal in the United States.

11. It is highly improbable that problems of the sort encountered in the behavior of Americans after capture in the Korean War can be offset by operation in terms of the armed force systems alone. The actual extent to which members of armed force organization can increase or lower their level of morale without reference to other social contexts, however, is nowhere thoroughly studied by modern scientific techniques as far as I know.
12. In relatively modernized contexts, armed force personnel may in fact be less isolated and are certainly less isolable than in relatively nonmodernized contexts, but interdependency in general is so high in the former that any isolation becomes more extreme in its implications—whether for good or for evil.
13. Prior to the beginning of World War II the situation of the armed force organizations of the United States illustrated this. Not only were there the ordinary cliques associated with relatively isolated membership units of any sort, but in addition the general civilian attitude can hardly be described as responsible. Civilians felt that armed forces should and could be called into being on a moment's notice and that professional military careers were hardly calculated to appeal to anyone of ability or sensitive tastes. Regular officers were looked down upon with the exception of rare personalities who had somehow come to prominence in time of crisis and held over some residue of reputation as the heroes of past wars. The attitude toward enlisted personnel who constituted the largest portion of the membership was even more extreme.

    Under these circumstances it was a minor miracle of American life that a sufficient skeletal organization of the armed forces in time of peace could be maintained to serve as a basis of rapid build-up in times of external emergency for essentially new armed force systems. The fact that there should be West Point and Annapolis cliques, that Army brats tended to marry other Army brats, etc., is by no means difficult to understand. The remarkable thing is that these tendencies did not go much further. With the end of World War II and the continuing crisis in world affairs following that period, the situation changed somewhat. Nevertheless, one of the strongest testimonies of the relevance of the general social setting for the character of the members of the armed force organizations is the fact that—despite the isolation inherent in armed force organizations as such, and the increased isolation by the general lack of public regard for armed force personnel in any time save times of crisis—the professional members of the armed force organizations did not come to consider themselves a breed apart and opposed to the general membership of United States society. It is difficult to understand why they did not do so, and their previous socialization is probably the only clue we have to such understanding.
14. Quite apart from the specific features of United States society, it may be argued generally that the relevance of armed force organiza-

tions as an instrument of external relations is more clearly appreciated than the relevance of armed force organizations as means of internal control. There are one or two cases of societies in the world which give the appearance of the *reductio ad absurdum* of the maintenance of armed force organizations for purposes of either internal or external control. The press of the world takes periodic pleasure in pointing to small countries whose annual military budgets consists of only a few pennies. Nevertheless, these are clearly not societies of any scale in the sense discussed here. They are actually quite clearly subsystems of some of the larger-scale societies whose associated territories are adjacent to theirs. As soon as observations depart from this set of extremely small nations and contemplate other quite small but not nearly so small nations, one contemplates a set of nations whose members guard their independence jealously. I doubt that the members of such societies as Denmark, Switzerland, etc., seriously believe that they could use their armed force organizations as protection against their larger neighbors, but the maintenance of armed force organizations oriented to such defensive operations does symbolize, at least, the determination of the members of those societies not to be easily engulfed. This symbolism increases the likelihood that their large neighbors agree in time of crisis on their neutrality or their protection.

5. One of the few instances of violence in the course of the modernization of Japan was occasioned by just such a fundamentalist reaction to proposals to modernize the armed forces. One of the major older feudal leaders, Saigo, Lord of Satsuma, revolted because he felt specifically that the new general conscript soldiers could not protect the integrity of Japanese society from threatening powers elsewhere in the world. When he finally received a convincing demonstration that ordinary peasant soldiers equipped with the modern rifles of that day could mow down fierce samurai warriors with their terrible long swords, he obligingly committed suicide to indicate his realization of error. His fearless peerless demonstration of good intentions was immediately recognized, and his reputation is forever enshrined as the Robert E. Lee of Japanese history. Military experts of relatively nonmodernized societies are not usually that difficult to convince.
6. As more sons and daughters survive, the relevance of any one of them to any one family organization either for his labor or for purposes of family continuity diminishes. Indeed, the structure of universal conscription may partially solve an otherwise embarrassing problem, the survival of children who were always wanted before the increased probability of survival.
7. Caution is warranted here. Zeal for the point of the relevance of armed force organizations for social mobility during the process of modernization should not obscure the point that throughout history generally these organizations have been ladders of mobility. Individual cases, e.g., Napoleon, have been spectacular.
8. This was notably the case on the history of the Japanese Army. Despite the closed class, feudal background of that society, at a time when it was difficult for a member of the United States Army to rise from enlisted status to officer status, the structures of the Japanese Army were far more highly universalistic.

19. I would be willing to give considerable odds that some day we shall have at hand data indicating that the extent of social mobility on predominantly universalistic grounds in the armed force organizations of modern Japan was far greater than the social mobility on such grounds within Japanese universities. If one compares the various cliques characteristic of Japanese society prior to World War II, i.e., the financial clique *(zaibatsu),* the military clique *(gumbatsu),* and the university or scholarly clique *(gakubatsu),* there is little doubt that predominantly universalistic criteria—that rational objective recognition of achievement—were much more characteristic of the military cliques than of the other two, academic biases to the contrary notwithstanding. Something of the same may also be true of United States society. Because of the general attitude of civilians toward armed force organizations, such organizations in the United States are not ordinarily thought of as a major ladder of social mobility despite the number of individuals who have risen to the presidency of the United States in part through their careers as important leaders of armed force organizations. Nevertheless, it would be interesting to compare social mobility in terms of armed force organizations and other general types in the society.
20. Some of these leaders maintain a sense of bravado by carrying pearl-handled side arms and the like, but it is quite clear that these are symbols of roles which the general can no longer play rather than essential equipment for roles that he actually plays. At all levels of modern military organization the premodern emphasis on individualism in battle which has always been so strong a focus of the romantic appeal of combat is preserved primarily in the naming of vehicles rather than in individual exploits. Contrary to clichés about regimentation, loss of personality, etc., prior to modernization few spheres featured such emphases on individualism as the military, but with occasional exceptions modern warfare is as much a matter of mass production as is modern factory work or modern large-scale farming.

2 *Political Development and Military Intervention in Latin America*

MARTIN C. NEEDLER

It is noteworthy that the recent spate of writings in the field of "political development" has shown a pronounced tendency to omit consideration of Latin America. Thus the "communications" and "bureaucracy" volumes in the SSRC political development series[1] are totally innocent of Latin American data, as is an excellent recent treatment of—of all things! —the political behavior of the military in developing areas.[2]

The Latin Americanists, for their part, have largely stressed those key features of the area's politics which have long remained constant—executive predominance, military intervention, and the influence of the peculiarities of Hispanic culture. At the same time, it is clear that the social changes usually collec-

Reprinted by permission of the author and the publisher from *American Political Science Review* LX (September 1966), pp. 616–626.

tively termed "modernization"—urbanization, technological borrowing, and the development of mass communications grids—together with their political correlate, the expansion of the political community to include hitherto excluded social elements, are proceeding in Latin America too.[3] Accordingly, it becomes desirable to reexamine the "statics" of Latin American politics in the light of the "dynamics" of the processes of political development and social mobilization.[4]

The present article attempts this reexamination with respect to the most characteristic feature of Latin American politics, the coup d'etat and the establishment of a de facto military government.[5]

A priori, mutually contradictory theses about the relations of the military coup to social development can be constructed—and indeed the literature on the subject abounds in such contradictory theses, evidence to support each of which is always available.[6] These hypotheses focus on whether military intervention in politics, represented most typically by the extra-constitutional seizure of power, is (a) increasing or decreasing, and (b) occurring primarily with the object of promoting socioeconomic change or of resisting it. Their starting points are the changes assumed to be going forward in the armed forces—the growth of professionalism, recruitment from a wider range of the population, greater influence from the United States, etc.

Now if evidence can be cited on either side of a proposition about developmental tendencies, it is clearly necessary to quantify these items of evidence along a time dimension; what is needed, accordingly, are empirical data giving the change in the frequencies of the occurrence of each of the contrasting possibilities over time.

The empirical questions we want answered, therefore, are:

1. Since the breakdown of early twentieth-century stability began the current period of change in Latin America, have coups d'etat become more or less frequent?

2. What changes have been occurring in the function of the coup in relation to changes taking place in the larger society?

3. What are the effects of changes in the Latin American military on the form, structure, and timing of the coup d'etat, and what political significance do these effects have?

INCIDENCE OF COUPS D'ETAT

One must first eschew the hopeless task of trying to account for coups d'etat that were not successful. The categories of coups that were aborted, suppressed, or abandoned melt into each other and into a host of other non-coup phenomena so as to defy accounting. At the same time, of course, since coups are after all illegal, they are matured under conditions of secrecy which make it inevitable that the unsuccessful projects for coups which become known about represent a highly biased sample. At the same time, an unsuccessful coup attempt may be the work of one or two atypical people; its occurrence does not necessarily say anything about the state of the polity as a whole, as does a successful coup.

During the thirty-year period from 1935 to 1964, there were 56 successful changes of government by extra-constitutional means in the twenty independent countries of Latin America. The frequency of their occurrence was as shown in Table 1. That is, the number of successful coups normally fluctuates be-

TABLE 1: *Frequency of Successful Coups d'Etat in Latin America, by Year, 1935–1964*

1935	1	1945	2	1955	2
1936	3	1946	2	1956	2
1937	3	1947	1	1957	1
1938	0	1948	4	1958	1
1939	0	1949	4	1959	1
1940	0	1950	1	1960	1
1941	1	1951	2	1961	2
1942	0	1952	2	1962	2
1943	2	1953	1	1963	4
1944	6	1954	3	1964	2

tween one and three per year. The clearly exceptional period was that from 1938 to 1942, during which only a single coup took place.

The first explanation which suggests itself is that these were years of recuperation from depression in which economic conditions were improving and the performance of government was likely to be regarded as satisfactory.[7] One would then hypothesize that a successful coup or revolt is less likely when economic conditions are improving.

A very rough test of this hypothesis can be made on the basis of figures for annual changes in real per capita product given in the UN Statistical Yearbooks for the 1947–1963 period.

It should be borne in mind, here and at subsequent points, that statistical data from Latin America leave much to be desired. It should also be noted that the data are not complete. However, data are available, for most of the years during that period, for ten countries in which coups d'etat occurred.

During 1947–1963, it is possible to assert, real per capita income figures for these countries showed a rise over the previous year's figure 87 times, a drop 39 times, and remained the same ten times. (These figures cannot be assumed to be typical of Latin America as a whole, it should be noted, since it is precisely the countries whose economies are likely not to be improving which do not report reliable economic statistics.) If coups d'etat occurred without relation to the state of the economy, one would then expect at least twice as many coups to have occurred in years which showed improvement as in those which showed deterioration, since there were more than twice as many "improvement years" as "deterioration years." However, that is not the case. Of the fifteen coups occurring during years for which the economic data are available, seven took place during years which showed an improvement, seven during years of deterioration, and one when no change was reported. The incomplete nature of the evidence should be stressed; in future years more complete calculations will doubtless be possible; other factors, not now identifiable, may be partly responsible; but the available data are consistent with the hypothesis postulated, that the overthrow of a government is more likely when economic conditions worsen.[8]

It seems reasonable, accordingly, to regard the years of low coup activity from 1938 to 1942 as due to the economic recovery of that period. Since 1949, a very slight secular trend in the reduction of the frequency of coups may be discernible. Since economic conditions are generally improving, although irregularly, this too might be expected on the basis of the same premise. Yet it should be remembered that variation in economic conditions can be held responsible for only a part of the variation in the frequency of coups, as the data discussed above also show.

Coups d'Etat and Social and Political Change

We turn now to the question of changes in the function of the coup in relation to social and political change. This question is extremely awkward to get at, since the origins of coups are often obscure, and the intentions of those staging them mixed. The author nevertheless believes it sound to explain the coup functionally rather than genetically, or in terms of factors external to the military rather than of internal characteristics of the military establishment, because of several considerations.

First, a military coup is not made by the military alone. Almost invariably, the conspirators are in touch with civilian politicians and respond to their advice, counting on their assistance in justifying the coup to public opinion and helping to run the country afterwards. This relationship not infrequently takes the form of a coup only reluctantly staged by the military at the insistence of civilian politicians, who appealed to the officers' patriotism, the historic role of the army in saving the country at its hour of need—of which national history doubtless affords many examples—and so on.[9] The chairman of one military junta which had outstayed its welcome spoke bitterly of some of its latter-day detractors "who used to cry at the doors of the barracks asking that the constitutional government be removed and even used to complain about the apathy of the military who did not want to act."[10]

Second, among the various conspirators, with their varying orientations and objectives, the position of those who can most count on outside support, whose own objectives are most in harmony with the aims of major outside forces, will be strengthened.

Third, the autonomy of the military decision to intervene may further be reduced by the fact that the political situation to which the military respond has been "engineered" by outside groups desiring intervention so as to trigger military predispositions in that direction. It is not unknown, for example, for right-wing activists to fake "Communist" terrorist attempts in order to help create an atmosphere conducive to military intervention.[11]

If the military coup is thus frequently called into play by the workings of the political system, what is its function in relation to social and economic change? Clearly, its purpose must increasingly be to thwart such change. This is so because the point of the coup is to prevent from happening what, it is assumed, would happen in its absence.[12]

Since social mobilization is proceeding, that is, constitutional presidents are likely to be responsive to social classes of progressively lower status as these enter the political arena by moving to the city or otherwise become mobilized. The policies of each successive constitutional president are thus likely, on balance, to constitute a greater threat to the status quo than those of his predecessor. This may be interpreted to the military by those trying to secure their intervention as a threat to the personal interests of military officers in the economy at large, as a challenge to the military in its role of preserver of domestic order, or, most likely, as a long-term threat to the special status and privileges, and even the continued existence, of the military institution.[13]

It thus seems probable that as social and economic development take place: (1) military intervention increasingly takes the form of an attempt by the possessing classes to maintain the status quo; (2) military intervention is increasingly directed against legally elected presidents heading constitutional regimes; (3) interventions increasingly occur to forestall the election and inauguration of reforming presidents; and (4) popular resistance to military intervention increases, resulting in greater likelihood that a military coup will lead to open fighting.

An analysis of the 56 successful insurrections[14] which occurred in the 20 countries of Latin America during the thirty-year period 1935–1964 appears to confirm each of these hypotheses, and thus to substantiate the argument made above. Table 2 gives the numbers and percentages of insurrections during each of the three decades of the period in which:

1. The reformation of the social and economic status quo was clearly a goal of the conspiratorial group; this shows a decrease.
2. A low level of violence (essentially a bloodless coup without streetfighting or other popular involvement) was maintained; this also decreases.
3. Constitutional rather than de facto governments were overthrown; this shows an increase.
4. The insurrection occurred during the 12 months prior to a scheduled presidential election, or in the four months immediately following; this likewise increased.

TABLE 2: *Statistics of Successful Insurrections, by Decades, 1935–1964*

	1935–1944		1945–1954		1955–1964	
	no.	%	no.	%	no.	%
Total	16	100	22	100	18	100
(1) Reformist	8	50	5	23	3	17
(2) Low in violence	13	81	15	68	6	33
(3) Overthrew constitutional governments	2	12	7	32	9	50
(4) Around election time	2	12	7	32	10	56

Dynamics of Coups d'Etat

Even if it be granted that the major determinants of the occurrence of a successful coup lie in the functioning of the total political system rather than in the internal dynamics of the military institution, those dynamics are of significance in such questions as the timing of the coup, and become especially important in determining the directions followed after the coup is successful and its leaders installed in government positions.

An examination of this problem must start from an appreciation of the fact that officers of the armed forces are not dominated by a single political viewpoint but hold a variety of political orientations. The correlates of these political orientations in personal characteristics have not as yet been systematically evaluated and weighed for the Latin American military, along the lines of Morris Janowitz' *The Professional Soldier.*[15] However, available evidence suggests that on top of a primary set of conditioning factors such as those which the American voting studies indicate are significant in party preference—that is, family tradition, social and economic level, and ethnic or other particularistic identification—is imposed a second set of factors peculiar to the military profession: rank, branch of service, occupational specialty, and career pattern. In a situation in which a coup d'etat becomes a possibility, ranking military officers are called on to develop policy positions on the question of the continuance in office of the president. The position each officer assumes will have two components, one based on attitudes toward the president's personal abilities, his programs, and the arrangements he is making for the succession; the other, partially independent of the first, reflecting the officer's views on the question of military intervention in politics in general.

The changes which have been taking place in Latin American armed forces in recent years suggest that the variety of political views represented within the military services has been on the increase, as the social origins from which officers are drawn have become less upper-class,[16] as the range of military technical specialities has been extended, and as the sheer size of military establishments has increased.[17] At the same time the increasing complexity of the governmental apparatus and the steady expansion of the proportion of the population which participates in politics, together with the technical improvement in the means of communication, have meant that a military coup needs itself to be more complex, to be more carefully planned, and to involve more people if it is to be successful. Because of heightened popular involvement in politics, a coup is also more likely to lead to open fighting rather than being accepted passively by an indifferent population. Given the range of political orientations within the military services, then, the task of the organizer of a successful coup d'etat is thus to build up a coalition of officers of a size and character adequate to execute the successful coup. The prime mover or movers in organizing the coup must therefore be engaged over a period of time in the process of building a coalition which will eventually exceed, in size and "weight"[18] the minimum necessary to insure success.

The originators of the conspiracy and the first to join it are those most opposed to the president and his policies, while other officers of different political orientations and a greater commitment to constitutional procedures have higher thresholds to interventionism. However, as time goes on, these thresholds will be reached for many officers as the tendency of the president's policies becomes clearer, as the country's situation, seen from their point of view, worsens, or as the succession problem becomes more acute with the approach of the end of the president's term.

It is of course possible that as time goes on the changes which take place in the situation are such as to reduce the degree of hostility to the president on the part of the organizers of the conspiracy, which may then disintegrate. It seems clear, however, that a successful coup would show a curve of support within the ranks of the military, rising over time and beginning with the original instigator of the plot, who represents the most extreme opposition to the president. The development of the curve of military support for the coup is likely to be exponential as the end of the president's term approaches. Under normal conditions the president prepares to hand over power to a successor of his own party or orientation, sometimes using not only his personal influence but also extralegal techniques to guarantee the succession. This raises the prospect of another four or six years of the same policies; yet the trepidation of those who oppose them necessarily increases. The heir apparent is in part an unknown quantity, which is disquieting; his previous public service will normally have taken place as a member of the president's cabinet, in which his own views necessarily had to be subordinated to those of his chief.

If there is a chance that the heir apparent would be defeated in the elections, the conspirators may await their outcome before striking. If he is indeed defeated, the need for conspiracy disappears; if he is elected, it then become necessary to strike before his inauguration, since his actual occupancy of the presidency would enable him to consolidate his power. Yet it is risky to wait until after the elections, which will mobilize his supporters and which may give him a strong mandate and thus strengthen his position with domestic and foreign opinion.

Thus, for these reasons also, the likelihood of a coup d'etat could be expected to increase as a president's term wears on, reaching its high point prior to a scheduled election but remaining high until the inauguration of a new president, this tendency becoming more marked over time in response not only to the accelerating social mobilization of the masses but also to the increases in the size, technical differentiation, and range of social origins of the officer corps.

Within the group of conspirators, then, a series of thresholds to interventionism is present, the lowest being that of the instigator (or group of instigators) of the plot, the highest being that of the last man (or group) to join in the coup before it is launched. The position of this hypothetical last adherent to the conspiracy is interesting to consider. If one recalls that the success of the coup is predicated on the formation of a decisive coalition to support it, then it is clear that the last adherent or set of adherents to the movement provided the critical margin of support, not just in its size, but especially in its "weight."

The importance of this hypothetical "swing man" in the situation, that is, may be due to any one of a series of factors—his personal influence within the armed forces, his prestige among the public, and/or his critical position in the command structure of the armed forces. It then becomes probable that because of his higher rank, greater prestige, and crucial importance for the coup, the "swing man" is placed at the head of the provisional government that emerges after the revolt is successful—as provisional president, as chairman of the ruling military junta, or as minister of the armed forces behind the facade of a civilian provisional government.[19]

An interesting and paradoxical situation is thus created. The "swing man" becomes the leading figure in the new government; yet he is the person who was least committed to the objectives of the coup, whose threshold to intervention was the highest of all the conspirators, and who was a last-minute addition to the conspiracy perhaps out of sympathy with, or not even aware of, the more fundamental aims of the group that hatched the original plan. Indeed, a situation can actually be created in which the head of the new government actually sympathized with the aims of the conspiracy not at all, but joined it at the last minute only to avoid pitting brother officers against each other, possibly precipitating a civil war.

These characteristics of the "swing man" can perhaps be made clearer by an illustration. A classical occupant of the role of "swing man" has been Marshal Castelo Branco of Brazil. A *New York Times* reporter described his position in the 1964 coup as follows:

> General Humberto de Alencar Castelo Branco has been called a "general's general." He rose to his present post of Army Chief

> of Staff after a long professional career in which he gained the high respect of his fellow officers but remained virtually unknown to the general public. . . .
>
> In the present crisis, the soft-spoken general first played the role of the reluctant dragon in refusing to join the developing movement against President Joao Goulart. His scruples were the same as those of many other Brazilian officers: The Brazilian Army has a tradition of protecting legality and the Constitution, and General Castelo Branco was not eager to become involved in a coup against a constitutional President.
>
> But the general became convinced that the continuation of the Goulart regime would lead Brazil to chaos and possibly a sharp shift toward the extreme left. He then drafted a position paper, the "Castelo Branco analysis," that became the justification for the army's support of last week's rebellion.
>
> Such is the respect enjoyed by the short, stocky, bull-necked general, that his analysis served as the turning point in the hesitations of many commanders in the crisis over Mr. Goulart.[20]

Clearly, in this kind of situation ample material exists for a conflict to emerge within the new provisional government. The conflict develops along the following lines. The erstwhile "swing man," now, let us say, president of the provisional junta, regards the objectives of the coup as realized with the overthrow of the former president and begins to make preparations to return the country to constitutional normality and to hold elections. The original instigator of the coup and the group around him, on the other hand, resist this tendency and instead urge the necessity for the military to keep power for a longer period, to purge all sympathizers with the deposed president completely from public life, to outlaw his party indefinitely, and to restructure political life to make it impossible for the tendency which he represents again to come to power.

During the recent period the basic situation described above has been reproduced in reality most faithfully in Argentina, Brazil, and Peru, and with local variations in Guatemala, Ecuador, the Dominican Republic, and Honduras.

In Argentina this basic set of dynamics has played itself out again and again since the overthrow of Perón in 1955, the irreconciliable anti-Perón forces being known as the *colorados,* or "reds," whose most characteristic figure is Admiral Isaac Rojas.[2]

Due to the more amorphous character of politics in Brazil, the same basic situation crystallized more slowly. The opposition between the military irreconcilables and the heirs of Getúlio Vargas has nevertheless been waged intermittently for ten years. The coup staged to prevent the inauguration of Kubitschek and Goulart in 1955 was unsuccessful; the coup designed to prevent the inauguration of Goulart as president in 1962 succeeded merely in having the powers of the presidency temporarily curtailed; only with the overthrow of Goulart in 1964 was the military anti-Getulista movement fully successful. After the successful revolt of 1964, the pattern described above became operative in its purest form, with conflict developing between the prestigious "swing man," Marshal Castelo Branco, metamorphosed into provisional president, and the *linha dura,* the "hard line" of the irreconcilable military opposition to the heirs of Vargas.[22]

A similar process took place in Peru following the coup d'etat of 1962. For 30 years the commanding officers of the armed forces had resisted the assumption of power by the revolutionary APRA movement, despite the fact that it commanded a majority, or at least a plurality, of the votes during the entire period. The party had begun in the 'twenties and 'thirties as a revolutionary Marxoid group, strongly anti-Yankee and prepared to use violence. During the thirty-odd years of its sojourn in the wilderness, however, the party leadership, and especially the party's founder, Victor Raúl Haya de la Torre, had "evolved" to a more moderate position, of which anticommunism was the central principle. At the same time, in the search for a respectability which would allay the misgivings of the military about the party, APRA's major tactician, Ramiro Prialé, led the party into alliance with increasingly more conservative forces, culminating shortly after the 1962 presidential election in an entente with the forces of General Manuel Odría. This was clearly the ultimate stage of the party's evolution, since Odría was a former military dictator who had outlawed and persecuted the party during his period of office, and who had run his election campaign in 1962 on a militantly anti-APRA platform. Haya had gathered more votes than any of the other candidates in the presidential elections of 1962, although only a handful more than Fernando Belaúnde and fewer than the one-third of the vote necessary to prevent the election's being decided by Congress. However, the new Congress, due to the vagaries of the electoral system, heavily overrepresented the APRA. Immediately following the election, the coup was staged, the leaders of the armed forces implausibly charging that the electoral results were vitiated by widespread fraud. In an unsuccessful last-minute attempt to avert the coup, the APRA leadership announced that its congressional votes would go to General Odría in a self-sacrificing attempt to break the impasse and avert the breakdown of constitutional procedures.

This situation made possible the emergence of a more muted version of the split which occurred in the Argentine and Brazilian cases. The ranking officer of the military junta, General Ricardo Pérez Godoy, was willing to return the country to constitutionality on the basis of the APRA offer to have its congressmen vote for Odría. The two key younger members of the junta, Generals Lindley López and Vargas Prada, who had personal and family ties to Belaúnde, opposed this solution, which would enable the APRA to exact concessions, for example in the shape of posts in an Odréa administration. Pérez Godoy was accordingly forced to resign and the reconstituted junta presided, during 1963, over elections in which, because of the withdrawal of two minor candidates, Belaúnde was successful.[23]

It appears overwhelmingly likely that as time goes on and popular participation in the processes of politics becomes greater, the Peruvian type of situation, in which over as long a period as necessary the popular choice is kept out of the presidency by repeated military intervention, will become increasingly common. As was suggested above, the pattern has extended itself to Argentina and Brazil already. Honduran politics seem to be moving in the same direction as the army has become increasingly committed against the Liberal party; the Guatemalan military staged their coup in 1964 to prevent the return to power of Juan José Arévalo; and the Dominican armed forces have clearly attempted to assume a similar position relative to Juan Bosch and the Dominican Revolutionary Party.

The logic of this type of situation suggests that the conflict between the most popular individual or party on the one hand, and the military irreconcilables on the other, tends to go on for some time, rather than being resolved by a single coup. This occurs for two reasons. In the military junta which forms after a coup, first of all, the irreconcilables normally are in a superior strategic position. The more moderate "swing man," whose prestige has entitled him to the chairmanship of the junta, may wish to restore constitutional processes as soon as possible. If this is likely to lead to the coming to power of the individual or tendency originally vetoed by the coup, however, the position of the junta president becomes untenable. Although he occupies the position with most authority and he may have placed close associates in the cabinet, theirs are not the key posts under showdown conditions: the key posts belong to those in direct command of troops, that is, the minister of the armed forces, the three service commanders, and even the commanders in the field. Because of this lack of congruence between the positions of authority when affairs are moving smoothly and the positions of power when a split develops, it is normally easier in such a situation to stage a coup d'etat than to prevent one.[24] The odds are therefore that the irreconcilables will be able to prevent the return to constitutionality for an extended period if this should seem likely to favor the archenemy.

Once military elements have vetoed the popular leader and his party, moreover, the hostility between the two becomes self-perpetuating and self-reinforcing, since those who participated in the original coup have reason to believe they will forfeit at least their careers, and perhaps more, if the outlawed party should ever gain power. As one Dominican colonel put it after the coup of 1963 when he was asked his attitude toward a return of Juan

Bosch: "If Bosch ever comes back, he will throw me into jail so deep I will never find my way out." Because of this set of circumstances the restoration of constitutional procedures becomes extremely difficult: unless the distribution of voter sentiment changes drastically, it is only too likely that the person or party which secured a majority in the last election would do as well in the next one. A temporary return to constitutionality may be possible on the basis of rigged or restricted elections, as has been the case in Argentina. Nevertheless, the Argentine political problem is not permanently solved. Given the persistence of the military irreconcilables and their point of view, the only permanent resolution of the problem lies in: (1) the definitive removal from the political scene of the vetoed leader by death or his renunciation of politics; (2) a shift in the distribution of popular opinion to the disadvantage of the vetoed party; or (3) the party's gaining respectability by drastic modification of its program or tactics. In Latin America, the third alternative seems a formal possibility only, since the irreconcilables may simply refuse to believe that the shift toward respectability is genuine. Thus the military veto against the APRA was still applied in 1962 despite the party's evolution to a moderate center or even right-of-center position. If the political problem has in fact been resolved in Peru —and this is not yet clear—it has been by way of the second alternative, in that the APRA may have been driven permanently below a third of the vote by a combination of disaffection from the left as the party's leadership has grown more conservative, and the permanent establishment in popular favor of Belaúnde's *Acción Popular*. Elsewhere a similar result may be achieved, at least temporarily, by the expansion or contraction of the electorate to shift the balance of forces against the vetoed political movement—by giving the vote to resident aliens, for example, or by taking it away from illiterates.

Conclusions

This examination of the internal logic of the Latin American coup d'etat in the circumstances of the current phase of history has so far led to three conclusions. First: the overthrow of a government is more likely when economic conditions are deteriorating. Second: as the military services have become larger and more various in the social origins of their officers, as military occupational specializations have become more differentiated and more highly professionalized, and as elections have become representative of the sentiments of a wider range of the population, coups d'etat have tended increasingly to occur in the period immediately prior to a presidential election and the subsequent inauguration, to be conservative in policy orientation, to be directed against constitutional governments, and to be accompanied by violence. Third: the tendency has emerged for conflict to develop, following a coup d'etat, between a more fundamentalist "hard line," and a "soft line" that shows greater readiness to restore constitutional procedures and is normally represented by officers of higher rank, occupying positions of greater prestige in the provisional government.

It is possible to draw a further conclusion, with policy implications for the United States, from this analysis. There has long existed a difference of opinion among students of U.S. foreign policy as to both the desirability and the feasibility of attempting to discourage military seizures of power in Latin America. The desirability argument is outside our present province,[26] but it is possible for us now to add something on the feasibility question —that is, how successful United States attempts to discourage military coups can be.

The failure of the United States to recognize a provisional government issuing from an extraconstitutional seizure of power, plus the imposition of other mild sanctions such as the suspension of military and economic aid, is of different effect to countries differently situated. The smaller countries whose economies are more dependent on actions of the United States—Bolivia, plus the countries of Central America and the Caribbean—are more susceptible to United States pressures than the larger South American countries. Nevertheless, examples can be cited of military coups which have taken place despite clear United States opposition, even in countries in the Caribbean area. These have been regarded as indicating that American opposition to such coups is ineffectual. The coup which took place in Peru in 1962, and the 1963 coups in Honduras and the Dominican Republic, for example, took place in the face of strong and explicit American opposition.

It still seems premature to conclude that American opposition to the military seizure of power is bound to be ineffectual, however. One problem here is methodological, since it is not possible to enumerate the coups d'etat that did *not* take place (although in two countries where American influence is heavy and which have known a history of military seizures of power, Venezuela and Panama, the constitutional succession has in recent years been unbroken while other countries of the area were experiencing violent changes of government). If the analysis made above is correct, however, the success of a coup d'etat depends, especially where the military services are large and highly differentiated, on the adherence to the coup in its later stages of officers least committed to its goals, less inclined to military intervention, and with more prestige and a higher position at stake. Since the success of the coup thus may well depend on its being joined by relatively few officers with a relatively weak commitment to its goals, it seems overwhelmingly likely that *any* deterrent to intervention—such as the suspension of military aid, or a credible threat not to recognize the new government—while not sufficient to deter the hard-core organizers of the coup, may nevertheless be sufficient to tip the scales against intervention for the crucial "swing man," or for the small group occupying the same tactical position, and thus may cause the coup to be abandoned, or to be launched without adequate support and thus to prove abortive.

In the coup situation, accordingly, even the mildest deterrent threat, such as a firmly stated nonrecognition policy on the part of the United States, may still be effective, because of the pre-coup balance of forces.

Light can also be thrown on the general problem raised in the opening section of the article, that of the relation between the constant features of Latin American politics and developmental trends, by an examination of the varying incidence over time of the characteristic military dictatorship in the area.

Here observers have tended to divide into the optimists, who perceive the evolutionary forces at work in the area as tending in a democratic direction, and the cynics, who take an attitude of "plus ça change. . . ." The incidence of military dictatorships in the area seems to support the views first of one group then of the other. A few years after Tad Szulc published his *Twilight of the Tyrants,*[26] which celebrates the replacement of dictators by democratic regimes, Edwin Lieuwen can write a *Generals versus Presidents,* which analyzes the reverse phenomenon.

The relation of cyclical and evolutionary patterns on this point can best be demonstrated by a graph. Figure 1 plots the number of unequivocally dictatorial regimes[27] in power during at least six months of each year over the period of the last 30 years during which evolutionary changes have clearly been occurring.[28]

Conclusions of great interest can be drawn. Clearly, the fac-

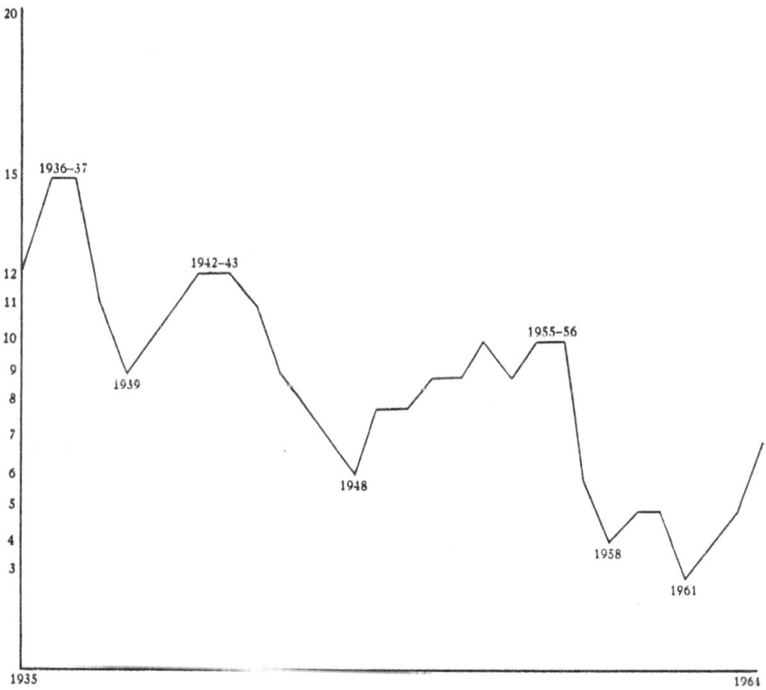

FIGURE 1: *Number of dictatorial governments in power in Latin America by year, 1935–1964*

tors which produce military dictatorships seem in part cyclical. At the same time, the cyclical pattern reproduces itself around a clearly descending trend line, so that each successive peak in the number of dictatorships existing contemporaneously is lower than the last: in 1936–1937, there were fifteen dictatorships; in 1942–1943, there were twelve; in 1955–1956, there were ten. Similarly, successively lower levels of dictatorship are reached at each low point of the cycle: in 1939, nine dictatorships; in 1948, six; in 1961, three.

The conclusion seems inescapable that, in this respect as in the others examined, while Latin American politics has certain abiding characteristics which produce its distinctive features, these are being progressively modified under the influence of forces of an evolutionary character.

NOTES

1. Lucian W. Pye, ed., *Communications and Political Development* (Princeton, N.J.: Princeton University Press, 1963); and Joseph La Palombara ed., *Bureaucracy and Political Development* (Princeton, N.J.: Princeton University Press, 1963).
2. Morris Janowitz, *The Military in the Political Development of New Nations* (Chicago: University of Chicago Press, 1964).
3. Karl W. Deutsch, "Social Mobilization and Political Development," *American Political Science Review* 55 (September 1961), 493–514; Bruce M. Russett *et al., World Handbook of Social and Political Indicators* (New Haven, Conn.: Yale University Press, 1964), especially pp. 294–298.
4. Two exceptions to the general lack of attempts to relate static and dynamic factors in a systematic way are the very fine article by Gino Germani and Kalman H. Silvert entitled "Politics, Social Structure, and Military Intervention in Latin America," *European Journal of Sociology* 2 (1961), 62–81; and Karl M. Schmitt and David Burks, *Evolution or Chaos: Dynamics of Latin American Government and Politics* (New York: Praeger, 1963).
5. This article forms part of a larger work currently in progress. I wish to thank the Horace H. Rackham Graduate School of the University of Michigan for the Faculty Summer Research Fellowship which enabled me to begin work on this subject, and the Harvard Center for International Affairs for the appointment as research associate which is enabling me to bring it to completion. I also wish to express my gratitude to Walter C. Soderlund, a doc-

toral candidate at the University of Michigan, whose research is reflected in the statistical data reported here.

6. A series of such pairs of mutually contradictory hypotheses drawn from the literature is neatly formulated by Lyle N. McAlister in his contribution to John J. Johnson, ed., *Continuity and Change in Latin America* (Stanford, Cal.: Stanford University Press, 1964), pp. 158–159. Some authors point out the evidence that various mutually opposed tendencies exist without attempting to subsume them in some general formulation. This is Johnson's own approach: see his *The Military and Society in Latin America* (Stanford, Cal.: Stanford University Press, 1964), Introduction and Chapter IX; and also that of Irving Horowitz, "United States Policy and the Latin American Military Establishment," *The Correspondent,* Autumn 1964. Lieuwen reconciles opposing tendencies by means of positing cycles in which a set of trends in one direction is succeeded by a counterveiling set; see his *Arms and Politics in Latin America,* rev. ed. (New York: Praeger, 1961), especially Chapter V.
7. Edwin Lieuwen discusses the relation between the depression and military intervention in "Militarism and Politics in Latin America," in John J. Johnson, ed., *The Role of the Military in Underdeveloped Countries* (Princeton, N.J.: Princeton University Press, 1962).
8. It should not be thought that economic conditions always worsen as a result of a coup. At least as commonly, in the writer's experience, conditions improve as business confidence shows an upsurge after a coup, which normally occurs without appreciable economic dislocation and typically removes a government regarded as incompetent.
9. McAlister argues in favor of this "revisionist" approach, which regards military intervention as chiefly a response to functional imperatives rather than as an expression of willful selfishness, in "Changing Concepts of the Role of the Military in Latin America," *The Annals* (July 1965), 90–92.
10. Admiral Ramón Castro Jijón, quoted in the *Diario Las Américas* (Miami), May 28, 1964. For a detailed account of the creation of an interventionist frame of mind on the part of the military, see Chapter V of my *Anatomy of a Coup d'Etat: Ecuador, 1963.* Johnson gives an example from Brazil of public incitement of the military to revolt by civilians on p. 124 of his *The Role of the Military in Developing Societies.* Samuel Finer discusses the interventionist mood in Chapter 5 of *The Man on Horseback* (London: Pall Mall, 1962).
11. For one such case of which the author has personal knowledge, see *Anatomy of a Coup d'Etat: Ecuador, 1963,* p. 19.
12. This is also Lieuwen's view: "On the balance, the armed forces have been a force for the preservation of the *status quo;* their political intervention has generally signified, as it does today, a conservative action. . . . " Edwin Lieuwen, "The Military: A Force for Continuity or Change," in John TePaske and Sydney N. Fisher, eds., *Explosive Forces in Latin America* (Columbus, Ohio: Ohio State University Press, 1964), p. 77.
13. It is the conclusion of Lieuwen's insightful *Generals versus Presidents* (New York: Praeger, 1964), pp. 101–107, that the last factor mentioned has been the crucial one in the recent coups.

14. Successful insurrections took place during the period in Argentina: June 1943, February 1944, September 1955, and March 1962; Bolivia: May 1936, July 1937, December 1943, July 1946, May 1951, April 1952, and November 1964; Brazil: October 1945, August 1954, November 1955, and April 1964; Columbia: June 1953 and May 1957; Costa Rica: March 1948; Cuba: March 1952 and January 1959; Dominican Republic: September 1963; Ecuador: August 1935, October 1937, May 1944, August 1947, November 1961, and July 1963; El Salvador: May 1944, October 1944, December 1948, October 1960, and January 1961; Guatemala: July 1944, October 1944, June 1954, and March 1963; Haiti: January 1946, May 1950, and December 1956; Honduras: October 1956 and October 1963; Nicaragua: June 1936; Panama: October 1941, November 1949, and May 1951; Paraguay: February 1936, August 1937, June 1948, January 1949, September 1949, May 1954; Peru: October 1949 and July 1962; Venezuela: October 1945, November 1948, and January 1958.
15. Morris Janowitz, *The Professional Soldier* (New York: The Free Press, 1964).
16. Medina Echevarría states flatly that the officers seizing power today "come, practically without exception, from hard-working middle-class families"; José Medina Echevarría and Benjamin Higgins, *Social Aspects of Economic Development in Latin America* UNESCO (Paris: 1963), Vol. II, p. 88.
17. The evidence for these developments is summarized by Edwin Lieuwen in Chapter 5 of *Arms and Politics in Latin America,* rev. ed. (New York: Praeger, 1961), pp. 122–153. If one thought solely in terms of these factors, as some authors do, regarding military political activity exclusively as being "pushed" by pressures internal to the military rather than being also "pulled" by the demands of the total political situation, then it would be logical to expect these changes to result in greater professionalism and technicism, reducing military involvement in politics, and in greater sympathy with the lower classes, rendering such involvement more progressive in orientation. Although several authors have assumed viewpoints of this type, they do not appear substantiated by the evidence cited above.
18. The concept of "weight" is discussed below.
19. This set of dynamics is of course not peculiar to Latin America. Classic occupants of the role of "swing man," with local variations, have been Naguib in Egypt, Gürsel in Turkey, Aguiyi-Ironsi in Nigeria—or even de Gaulle in France.
20. "Man in the News," *New York Times,* April 6, 1964. Typographic errors in the original have been corrected.
21. See Arthur Whitaker, *Argentine Upheaval* (New York: Praeger, 1956); Lieuwen, *Generals versus Presidents,* pp. 10–25; James W. Rowe, *The Argentine Elections of 1963: An Analysis* (Washington: Institute for the Comparative Study of Political Systems, n.d.), pp. 11–18; Peter G. Snow, "Parties and Politics in Argentina: The Elections of 1962 and 1963," *Midwest Journal of Political Science* 9 (February 1965), 1–36.

22. See *Generals versus Presidents,* pp. 69–85; and Phyllis Peterson, "Brazil: Institutionalized Confusion," in Martin C. Needler, ed., *Political Systems of Latin America* (Princeton, N.J.: Van Nostrand, 1964), pp. 473–477.
23. See *Generals versus Presidents,* pp. 26–36; and M. C. Needler, "Peru Since the *Coup d'Etat,*" The World Today, February 1963.
24. In one variant of this situation, the provisional president may save his own personal position by switching sides at the last minute and adopting the program of the "hard liners," if the forces they can marshal seem decisive. This tactic was adopted by Castelo Branco in early 1966.
25. I have discussed it in "United States Recognition Policy and the Peruvian Case," *Inter-American Economic Affairs* (Spring 1963).
26. (New York: Holt, Rinehart and Winston, 1959).
27. To be considered "dictatorial," a government: (1) Had to be not an avowedly provisional regime holding office for 36 months or less; (2) Had to come to power, or remain in power after the conclusion of the constitutionally prescribed term of office, by means other than a free and competitive election; or rule in clear disregard of constitutionally guaranteed liberties.
28. The idea of approaching the problem in this fashion was suggested to the author by Ronald Schneider's article "The U.S. in Latin America" in *Current History* (January 1965).

3 The Military and Political Change in Asia

JOHN P. LOVELL
C. I. EUGENE KIM

A cursory review of recent developments in a number of Asian states reveals a pattern of vast and significant influence by military elites upon the course of political change. In Indonesia, General Suharto, backed by the army, has assumed all executive powers, climaxing a consolidation of power that began following the Untung Coup in 1965. In Communist China, the army has been a mainstay of the dictatorship of Mao Tse-tung. In Pakistan, despite some restoration of parliamentary procedure, General Ayub Khan remains in control of the government, as he has been since the military coup d'etat in 1958. In Burma, the military, led by General Ne Win, exerts dominance not only over the formal institutions of government but also over many

Reprinted by permission of the authors and the publisher from *Pacific Affairs* XL (Spring-Summer 1967), pp. 113–123.

other sectors of society as well. In South Korea, outright military rule, which began in 1961, has given way to a pattern of "quasi-civilianization," in which political power remains anchored largely in the military, and the vast military establishment continues to exert a profound influence upon Korean society. In South Viet Nam, a proposed transition from military to civilian rule may soon be a reality, but the military will participate extensively in the governing process. In Thailand, a long tradition of military dominance of the political process continues. Even though the Philippines is characterized by civil supremacy in government, the influence of the military extends beyond its role of defense into areas such as internal security and "civic action." In India, a largely British-trained elite of military officers displays a deep commitment to the norms of civil control of the military, but continuing domestic unrest and governmental instability, or renewed external threat, could conceivably provoke the Indian military elite into employing the enormous political potential represented by an organization of nearly one million men for the purpose of assuming control of the political process.

A growing body of empirical and theoretical analyses of the role of the military in developing nations provides some insight into the pattern of military influence in Asian politics.[1] On the other hand, important gaps in our knowledge of the relationship of the military to political change remain. This article examines the state of our knowledge in the social sciences of the role of the military in politics, with emphasis on the important gaps and why they merit attention in future research.

Consistent with the general increase of systematic attention in the social sciences in recent years to Asia and other non-Western areas, there has been a gradual erosion of the earlier view that major participation by military groups in political change is necessarily perverse. As Guy J. Pauker has stated, "our liberal tradition [made] it repugnant to contemplate regimes controlled by military elements rather than by the civilian instrumentalities of representative government. . . ."[2] Increasingly, however, social scientists have been willing to consider the consequences of military participation in political change in an empirical manner, and in a growing number of studies the role of the military in developing nations is viewed from an objective analytic perspective. These studies have been primarily of two varieties. Most numerous have been individual case studies. More limited in number but broader in scope have been studies synthesizing observations of a number of political systems in order to formulate general propositions or middle-range theories about the political role of the military. Examples of the latter variety of study include those of Andrzejewski, Pye, Finer, Janowitz, Lieuwen, Johnson, and Gutteridge.[4]

The focus of much of this systematic research on the military in developing nations has been the military coup d'etat. One finds most often three key sets of independent variables to explain this phenomenon.

First is a set of variables relating to the *susceptibility* of the political system to the forcible seizure of power. Included are variables which refer to the general pattern of legitimacy within the society attaching to existing institutions of government and defining the appropriate means by which authority within the system may be transferred; Finer's concept of the "level of political culture" within a society is of this general nature. In his book, he discusses countries of "developed," "low," and "minimal" political culture.[4] Other studies have emphasized variables which refer to broad and fundamental transformations of the system producing instability and thereby placing a greater premium on the use of force in politics. Lieuwen notes that a change of this sort occurred in Latin America following the onset of world depression in 1930, leading directly to a pattern of military dominance in politics.[5] Or, variables related to susceptibility of the system to forcible seizure of power may refer to specific phenomena more limited in scope or time, such as the emergency of an external threat to the system, or an event which discredits a civilian regime, or removes from the scene key elements of civilian control. For example, in South Korea the military coup of May 1961 followed the April 1960 student movement which toppled the First Republican regime of Syngman Rhee (1948–1960) and the ineffective Chang regime of the Second Republic (1960–1961). There had been a long series of negative evaluations of civilian politicians by the people. In Burma, a threat of civil war coupled with the chronic inefficiency of civilian government seems to have been the key factor which prompted in 1958 and also in 1962 the military to take over the government from U Nu, who was still a popular figure in Burmese politics at the time.[6] In Pakistan, the death of the revered Mohammed Ali Jinnah in 1948, followed by the death of his protégé, Liaquat Ali Khan, in 1951, were events which deprived the civilian government of their key leaders; these losses contributed to growing discontent and instability, culminating in a military coup in 1958. In Indonesia, even though the exact circumstances which surrounded the coup and the counter-coup in September 1965 are still shrouded in mystery, it is clear that the army's assumption of power was in part designed to break the paralysis of the country under the flamboyant leadership of Sukarno. Under growing economic distress in the country and disenchantment with the central government in Java, Sukarno was no longer able to maintain the precarious balance between and among various power-seeking institutions, the government, the army, and the Communist Party (PKI) being the dominant ones.[7]

A second important set of variables is that relating to the *capacity of the military* to execute a coup d'etat. In most of the states of Asia, the size alone of the military offers one crude index of the capacity of this group for political influence. Ironically, in some cases the eventual dominance of politics by the military was facilitated by decisions of civilian leadership to increase the size of the armed forces in order to cope with real or imagined external threats or internal unrest. In both South Korea and in South Viet Nam, for example, civilian policies to increase substantially the armed forces, implemented through the largesse of American aid, contributed directly to the emergence of the indigenous military as the dominant group in politics.

The reasons for the increasing dominance of the military in Asia are more complex than size alone. As Pye has aptly noted, "modern armies are somewhat easier to create in transistional societies than are most other forms of modern social structures."[8] Indeed, the development of military establishments with modern skills, equipment, and weapons is a characteristic of the transitional societies of Asia. Such skills, equipment, and weapons—along with size—comprise what Robert Dahl terms "political resources." Coupled with other resources prevalent among military establishments of contemporary Asia, such as relatively disciplined, cohesive organization, their political resources have given the military a capacity for forcible intervention in the politics of virtually every nation-state in Asia. Nevertheless, there is not a direct link between the political behavior of the military and their political capability. Rather, in order to explain their political behavior one must examine also their political attitudes and beliefs. Thus, a third set of variables identified in recent research relates to such attitudes and beliefs—the *political perspectives* of the military.

Salient among these perspectives are the images which they develop of their own appropriate role and status within the society, and their perceptions of civilian elites and of the political system as a whole. Under conditions where civilian political leadership is perceived by the military as corrupt, incompetent, or otherwise incapable of coping with the basic tasks of government, the motivation among the military for performing these tasks themselves directly increases. Evidence suggests that such perceptions were among those contributing to the decision of Ayub Khan and his military followers to take control of the government of Pakistan in 1958, the decision of General Ne Win and his officers to seize power in Burma the same year, and the decision of General Park Chung Hee and others to execute a coup in South Korea in 1960 (a plan postponed for roughly a year when the so-called Student Revolution brought about the downfall of Syngman Rhee).[9] Similarly, if civilian political leaders take action, or seem about to take action, detrimental to the personal or organizational interests of military men, the motivation to intervene in politics to neutralize or ward off the threat to their interests is likely to develop. For example, the announced policy of the Chang Myon government of South Korea to bring about substantial reductions in the armed forces was one of several factors which activated those within the military (Park Chung Hee, Kim Chong P'il, and others) who brought about the overthrow of the Chang Myon regime.[10] In Indonesia, Sukarno's support for the Communist-supported plan of arming a militia force of peasants, who would serve as a counter-force to the regular army, seems to have contributed to the dissatisfaction which led to the Untung coup in 1965.[11]

Nevertheless, in spite of abundant similar examples, it is important also to note a number of instances where military men have had a capability for forcible intervention in politics but have refrained from acting. The cultivation of a sense of professionalism among military officers ordinarily includes norms which require the subordination of the military to the civil sector of government. In an environment where military institutions have become relatively developed, military leaders are likely to regard with reluctance possibilities for seizing control of the government on the grounds that active participation in politics is detrimental to professionalism. For example, in 1954 General Mohammed Ayub Khan declined the invitation of the Governor-General of Pakistan to take over the government. (One might add that Ayub apparently was tormented by doubt about the wisdom of intervention in 1958 before finally deciding to act.) The training of Ayub and other Pakistan military leaders in the British tradition of civil supremacy in government may have caused their reluctance. This is not to exaggerate the immunity of military men, even when highly professionalized, to political ambitions, but only to suggest that even when capable of seizing total political power, military leaders often are willing to accept less. The military of Indonesia, to cite another case in point, were willing to engage in a precarious form of political bargaining and competition with other groups (most notably the Communist

Party) until 1965, in spite of possession prior to that time of political resources probably sufficient for the successful execution of a coup.[12]

One notes also that if the idea of military control of government is sufficiently repugnant to important sectors within military leadership or within the society as a whole, even if the military are provoked into executing a coup, the lack of long-run justification for military control (even if a short-run rationalization for it is formulated) exerts pressures on the military regime to shift at least to a basis of "quasi-civilianization."[13] The elections in South Korea in 1963 and in South Viet Nam in 1967 were designed to provide such a transistion from outright military rule to "quasi-civilian" rule. But as the Burma case (in 1958) illustrates, although respect for civil rule may lead to its restoration after an interim period of "caretaker" rule by the military, a precedent has been established making it all the easier for the military to justify their intervening again (in 1962) the next time civilian leadership falters.

Undoubtedly, the coup d'etat is the most dramatic form of participation by the military in politics. Yet, seizure of the reins of government is by no means the only, or the most important, way in which military men and military groups influence politics. Perhaps more important in terms of long-run political change in Asia and in several other parts of the world are the roles which military groups play in the functions sometimes termed political "inputs"—political recruitment, political socialization, political communication, the articulation and aggregation of political interests and demands. These roles remain largely neglected by students of political development in general and of the role of the military in politics in particular.

Especially interesting to the student of political change and worthy of systematic investigation is the relationship of indigenous military elites to the processes of political socialization (that is, the process of acquiring political norms and beliefs) and political communication. "Modernization," after all, in part means the rejection or repudiation of some (not necessarily all) old values and beliefs and the acceptance or adoption of new ones.

Most of the new values and beliefs characteristic of transistional societies in Asia have been imposed or imported from abroad, from the West. Metaphorically, one can postulate two networks of communication through which these values and beliefs flow. The first is an international network linking a given nation to its external environment. Second is an intrasocietal network feeding communications internally to various parts of a society. Nearly all indigenous military establishments in Asia currently have a tremendous capacity to affect political change because of the key positions they occupy in this dual network of political communications.

To pursue the metaphor, the military may be viewed as one of the key mechanisms which a nation possesses for receiving, and sometimes for amplifying, signals from its external environment —these signals include ideas, values, skills, techniques, and strategies of political change. Like certain other professional groups, the military develop a "language" that cuts across national boundaries and links them with some members of the same profession in other nations. But more so than most groups, military men in many of the countries of Asia have been compelled to develop their facility for using this cross-cultural professional "language." An obvious case in point is the great number of Asian military men who come to the United States for instruction at various American institutions. This kind of direct contact with the American environment and American ideas has been at a rate of some 8,000 foreign military personnel per year in recent years. More than 90 countries have sent military personnel to the United States, South Korea and South Viet Nam having been particularly involved in these programs. Furthermore, in their own countries, indigenous military personnel in large numbers work side by side with American or other foreign military personnel. Again, South Viet Nam and South Korea are prominent examples; one could add Taiwan, Thailand, and the Philippines.

The Soviet Union has been another important source of modern ideas and techniques for Asian military elites operating in the international network of communications. The North Korean military was from its inception created with Soviet aid and under its leadership.[14] In Communist China, with its unique history of revolutionary war, Mao's sprawling semiguerrilla army was rapidly transformed into a modern army aided (until 1960) by the Soviet Union.[15] It is interesting that modernization of the military both in China and in North Korea has led to the development of an outlook among military professionals different from and often in conflict with the irregular, revolutionary, and "democratic" character of the insurrectionary officers of these states. Dramatic revelations of such conflicts were provided by the 1956–1958 period of the great purge in the North Korean army and the purge in the Chinese army leadership in 1958.[16]

Indigenous military men also transmit signals in an intrasocietal network. They do this through contacts with their own personnel and with civilians. One means of intrasocietal communication is provided by the education, training, and indoctrination which the military provide their own personnel. In Communist states, such as China, North Viet Nam, or North Korea, where political doctrine is highly developed and pervades all forms of education, the political relevance of education and training activities within the military is perhaps obvious.[17] Less obvious, perhaps, but meriting considerable further investigation is the political relevance of military education, training, and indoctrination in the non-Communist systems of Asia. The system of military education in South Korea offers one example of both the politically relevant substance of military education and the numbers of persons in the society exposed to such education.[18]

In addition to their exposure to basic military training, more than 300,000 Koreans have graduated by 1967 from one or more of the schools and courses within the military educational system of South Korea. Patterned largely after American counterparts, Korean military schools range in focus from those offering specialized training in a particular branch of service or technical specialty to those such as the service academies, the command and staff colleges, and the National War College, which, in preparing officers for top command positions, expose students to a wide range of domestic and international problems and issues. Particularly in the case of the latter, generalized leadership courses, but to some extent even in specialized courses, the military establishment through its educational program attempts to impart a common body of ideals and concepts, ranging from Marxism (taught from the standpoint of "knowing your enemy") to the goals of the Korean Military Revolution of 1961. As in the American military establishment, a cadre of Korean military officers and enlisted men are given special training and assigned particular responsibilities for carrying the values and ideas taught within the formal military educational system back to the unit level. Two hours of "troop information and education" instruction are prescribed for each unit per week.

One must emphasize also that aspects of military education and training other than formal indoctrination acquire political relevance to the extent that new skills or new beliefs become transformed into demands upon the political system or supports which maintain the system.

Because systematically collected data are not yet in a form where findings can be discussed, only anecdotal speculation can be offered about the impact of military education, training, and indoctrination.[10] Nevertheless, what one can note, at least in general terms, is the pivotal position of the military educational structure within the intrasocietal communications network. Not only have several hundred thousand Koreans been exposed to skills, ideas, and values stemming from a common institutional origin, but thousands of other Koreans have been exposed to the changed or reinforced perspectives of the military men. Military education and indoctrination, in other words, would seem to have an important cumulative impact. The man returning to his village upon completion of his military service brings with him new ideas about mechanical skills, the planting and harvesting of rice, and about the responsibilities of government to the farmer. Or, instead of returning to the village, an ex-serviceman seeks employment in the city, representing a trend of migration from rural to urban areas that has political (as well as social and economic) consquences for both village and city. One might cite other examples, but the point is that both in positive and in disruptive form, military experience has cumulative and far-reaching political implications in a transitional society.

Still another related form of intrasocietal communications in which the military loom large is what might be termed "symbolic communication." To the extent that they embody the symbols of partiotism and national independence, various facets of their behavior and outlook are likely to be emulated within the civilian sector of society. The most striking example is perhaps the pre-1945 Japanese case, where *bushido,* or "the way of the warrior," which had its origin in Japanese military history, was institutionalized not only as a set of ideals for the military class but also for every Japanese to follow, and the army indeed became the center of ultra-nationalist movements in prewar Japan. Even in contemporary Japan, where the military remain relatively small (and in some sectors discredited) since the Second World War, a number of private companies have required that new male employees spend a few days in an army camp just to observe what life is like in the armed services. A personnel training director for one of the companies participating in this project is quoted as seeing as the probable result of this program the "better appreciation of discipline, considerate behavior, and team work" by the young men through their contacts with the military.[20] A still more institutionalized form of emulation by civilians of the behavior and attitudes of the military has been enforced in China where, beginning early in 1964, a campaign was launched by the government to encourage all sectors of the society to view the People's Liberation Army as a model to "learn from, study, and compare with."[21]

Often the role of the military as an agent for influencing the attitudes and behavior of civilians extends beyond that of merely serving as a model and includes planned activities of indoctrination and training. In the Philippines and in Indonesia, the government utilized the military not only in the establishment of law and order but also in local improvements and development of virgin areas by settlement. One of the advantages of the military in this connection is the ability of the military "to send highly qualified specialists to areas for which it would be extremely difficult to recruit qualified civilians."[22] Following the Korean war, the Korean forces worked jointly with the U.S. Forces in the Armed Forces Assistance to Korea program in rehabilitating communities and in other "civic action" projects. The program employed by the Philippines to defeat the Huks during the years 1950–1953 is also a case in point. Among the key features of President Magsaysay's campaign were "military civic action, particularly village level economic development and social improvement."[23] In Burma, vast sectors of the economy have been nationalized and placed under direct or indirect military control since the second military takeover in 1962. Beyond this, the military in Burma participate in a variant of civil action activities through the Defense Service Resettlement Program, which provides for the resettlement of retired servicemen in a group of model villages in a highland area 300 miles northeast of Rangoon. A central administrative village, directed by an army officer, serves to coordinate administrative services to the various villages in the settlement and provides a central market, seed farms, a tractor station, a hospital, and other public services.

This type of civic action program has also been used very widely by the Communist regimes. As systems of total mobilization, Communist China, North Korea, and North Viet Nam have been mobilizing every sector of society for construction for socialism. This practice has meant not only the close merging of political and military doctrine in an organic whole but also subordinating the development of military professionalism to the duty of organizing the militia and participating in economic construction work.[24] The Chinese Red Army units grow their own foodstuffs and work in irrigation and other projects. They help with the harvest and, particularly in recent years, they often spend more time on productive labor than on field training.[25]

In summary, the indigenous military establishments of most of the nation-states of Asia are engaged in activities which place them in vital positions in relationship both to international communications and to intrasocietal communications. Their access and receptivity, in most cases, to ideas that are "modern," coupled with their contacts with and influence upon broad sectors of their societies from the "grass roots" level up, means that the military have a tremendous potential for imparting or shaping the basic political norms and beliefs affecting political change. This is not to say that the role of the military as agent of political change is necessarily good, any more than modernization is necessarily good. Indeed, the point to stress is that far too little is known about the social and political consequences of various military activities such as those described. We lack even adequate descriptive information about these activities which would be useful for the formulation of meaningful hypotheses.

Some significant generalizations have been produced about the most dramatic form of involvement by the military in politics in developing nations—the coup d'etat. Now we need to concentrate on the various forms of involvement by the military that are less dramatic but may substantially affect the course of political change in Asia and elsewhere in coming decades.

NOTES

1. See, for example, the collection of essays dealing with politics and the military in new states in Wilson C. McWilliams, ed., *Garrisons and Government,* (San Francisco, 1967).
2. Guy J. Pauker, "South Asia as a Problem Area in the Next Decade," *World Politics,* XI, no. 3 (April 1959), 343.
3. Stanislaw Andrzejewski, *Military Organization and Society* (London, 1954); Lucian Pye, *Armies in the Process of Political Modernization* (Cambridge, Mass., 1959); S. E. Finer, *The Man on Horseback* (New York, 1962); Morris Janowitz, *The Military in the Political Development of New Nations,* (Chicago and London, 1964); Edwin Lieuwen, *Generals versus Presidents* (New York and London, 1964); John J. Johnson, ed., *The Military and Society in Latin America* (Stanford, California, 1964); William F. Gutteridge, *Military Institutions and Power in the New States* (New York and London, 1965).
4. Finer, *Man on Horseback,* chap. 7, 8, and 9.

5. Edwin Lieuwen, *Arms and Politics in Latin America,* (New York, 1960), chap. 3, *passim.*
6. Louis J. Walinsky, "The Rise and Fall of U Nu," *Pacific Affairs* XXXVIII, nos. 3 and 4 (Fall and Winter 1965–1966), 269–281.
7. W. F. Wertheim, "Indonesia Before and After the Untung Coup," *Pacific Affairs* XXXIX, nos. 1 and 2 (Spring and Summer 1966), 115–127.
8. Pye, *Armies.*
9. C. I. Eugene Kim, "The South Korean Military Coup of May 1961: Its Causes and the Social Characteristics of Its Leaders" (a paper delivered at the Sixth World Congress of Sociology held at Evian, France in September 1966) to appear in a forthcoming volume, *The Armed Forces and Society: Sociological Essays,* Jacques Van Doorn, ed. (The Hague: Mouton, 1968).
10. *Ibid.*
11. See Frederick Bunnell, "Indonesia's Quasi-Military Regime," *Current History* 52 (January 1967), 22–28, 51.
12. See Guy J. Pauker, "Indonesia: The Year of Transition," *Asian Survey,* VII (February 1967), 141. Also Daniel S. Lev, "The Political Role of the Army in Indonesia," *Pacific Affairs* XXXVI (Winter 1963–1964), 349–364.
13. The term "quasi-civilianization" is Finer's; see his discussion in *Man on Horseback.*
14. Kiwon Chung, "The North Korean People's Army and the Party," *China Quarterly,* no. 14 (April-June 1963), 105–124.
15. Ellis Joffe, "The Conflict between Old and New in the Chinese Army," *China Quarterly,* no. 18 (April-June 1964), 118–140, particularly p. 122.
16. In this respect, the *Bulletin of Activities of the People's Liberation Army* is most revealing. See J. Chester Cheng, trans., *The Politics of the Chinese Red Army: A Translation of the Bulletin of Activities of the People's Liberation Army* (Stanford, Cal.: Hoover Institution, 1966).
17. For a brief discussion of the political indoctrination of officers in China, see Davis B. Bobrow, "The Good Officer: Definition and Training," *China Quarterly,* no. 18 (April-June 1964), 141–152.
18. The Korean experience is described and assessed in greater detail by John P. Lovell in "The Military as an Instrument for Political Development in South Korea," a paper presented at the Conference on Korea sponsored by the Committee on Asian Studies, Institute of International and Area Studies, Western Michigan University, April 1967. The proceedings of the conference are to be published in 1968.
19. In conjunction with Professor Young Ho Lee, University of Georgia, and scholars in Korea, John P. Lovell has administered an attitude survey throughout much of the ROK military educational system. Data from the survey are presently being analyzed. Preliminary reports of findings will be made late in 1967.
20. "Army Helps Train Workers in Japan," *New York Times,* March 13, 1966, p. 9.
21. John Gittings, "The 'Learn from the Army' Campaign," *China Quarterly,* no. 18 (April-June 1964), 153–159. Obviously, as Gittings points

out, such a program encouraging emulation of the PLA rested upon an assumption of firm loyalty and commitment of the PLA to the person and ideology of Mao Tse-tung. The recent struggle within China casts doubt upon that assumption; evidence suggests that the loyalties of the PLA have been divided.

22. See the chapter on "Military Civic Action" prepared by the Civil Affairs Directorate, Deputy Chief of Staff for Military Operations, Department of the Army, and included in Richard M. Leighton and Ralph Sanders, eds., *New Dimensions in the Cold War* (Washington, D.C.: Industrial College of the Armed Forces, 1963), pp. 173–182, particularly pp. 178–179.
23. *Ibid.*, p. 179.
24. John Gittings, "China's Militia," *China Quarterly*, no. 18 (April-June 1964), 100–117.
25. Edgar O'Ballance, *The Red Army of China* (New York, 1963), p. 214. See also Ellis Joffe, "The Communist Party and the Army," *Contemporary China* IV (1961), 55–59.

4 *Swords and Ploughshares: The Turkish Army as a Modernizing Force*

DANIEL LERNER
RICHARD D. ROBINSON

Military predominance in public life has a long and strong tradition in Turkey. Lybyer has told us: "The Ottoman government had been an army before it was anything else. . . . In fact, Army and Government were one. War was the external purpose, Government the internal purpose, of one institution, composed of one body of men."[1] And Gibb quotes el-Gazali, at the end of the eleventh century: "Government in these days is a consequence solely of military power, and whosoever he may be to whom the holder of military power gives his allegiance, that person is the caliph."[2]

Reprinted by permission of the publisher from *World Politics* XIII (October 1960), pp. 19–44.

These themes are still resonant when one considers the founding of the Turkish Republic. The "one body of men" who seized and shaped the new republican power was a military body. The loyalty of this elite, as of the larger populace, was given to the Gazi—the triumphant military commander. Mustafa Kemal Ataturk was, one recalls, the only Turkish officer to emerge untainted by defeat throughout the disastrous period that began with the Balkan wars and ended with the Greek invasion. It was the military corps that named, and military prestige that sustained, the leader—once a sultan-caliph, now a president.

Once installed, however, Ataturk moved in a radically new direction. After the peace was signed in 1923, he obliged the leaders of the revolutionary government to divest themselves of military office. His own official portrait was taken in white tie and tails rather than military regalia. The man's uncanny feel for the vivid symbol revealed, in this case, a determination to civilianize the Turkish Republic that cut wide and deep.

Ataturk's position on this point was made abundantly clear as early as 1909 when he, along with Ismet Inonu (his successor as president and confidant throughout), argued vigorously in the Committee of Union and Progress against the direct participation of army officers in political affairs. At the annual party meeting in Salonika in the summer of 1909, Ataturk declared: "As long as officers remain in the Party we shall neither build a strong Party nor a strong Army. In the Third Army most of the officers are also members of the Party and the Third Army cannot be called first-class. Furthermore, the Party receiving its strength from the Army will never appeal to the nation. Let us resolve here and now that all officers wishing to remain in the Party must resign from the Army. We must also adopt a law forbidding all future officers having political affiliations. . . ."[3] On another occasion, that same year, Ataturk commented publicly, "To be victorious in the internal affairs of a country is due less to an army than to the successful offices of a government."[4] Shortly thereafter, both Ataturk and Inonu resigned from the party and, from 1909 to 1919, both men devoted themselves exclusively to military activity.

This policy of "civilianization" of administration, though not proclaimed explicitly as one of the six cardinal principles of the Republic, was nonetheless firmly imbedded in the Turkish Constitution of 1923, Article 40 of which clearly gave charge of the army to the Grand National Assembly and, as its representative, to the president. Also, it was declared unconstitutional for Assembly members to hold a second government office, presumably including army assignment. Likewise, the Constitution forbade the government to declare martial law for a period longer than a month without specific Assembly approval. The chief architects of the 1923 Constitution were Ataturk and Inonu.

In October 1924, anticipating an alleged military plot against his political authority, Ataturk requested all of his chief military commanders to resign their Assembly seats. He observed: "I have come to the conclusion that, for the maintenance of army discipline in required measure for the exercise of command, it is incompatible that commanders should at the same time be deputies."[5] Although the chief of the general staff was excluded in 1924 from the cabinet, of which he had been a member since its inception in 1920, he was nonetheless still answerable only to the president of the Republic. This situation continued up to the end of 1943, when, apparently in anticipation of Marshal Fevzi Cakmak's retirement, the army was placed more directly under full civilian control by making the general staff chief answerable to the cabinet and prime minister. Later, in 1949, a Supreme Council of National Defense was formed, on which sat several civilian cabinet ministers. This organ has in part balanced the power of the general staff.[6] The election laws of Turkey now go so far as to disfranchise all military officers, soldiers, military students, and members of the police force. They may not vote in any election, or enter as candidates, or be politically active.[7]

In retrospect, it is quite evident that Ataturk deliberately attempted to fence the military off from active political life as soon as its political support was no longer essential to his power. Even the two high-ranking army officers who, following Ataturk's 1924 edict, had resigned their military offices to retain Assembly seats were soon deliberately discredited politically by Ataturk.

In short, Ataturk undermined the central proposition of the classic Persian *Mirror for Princes* literature, which had been domesticated in the Ottoman imperium over the course of centuries. Gibb has formulated this proposition concisely: "The prime object of the ruler's munificence must be his army: 'the king's enemy is his treasury and his friend is the army; when one is strong, the other is weak.' " Flying in the face of this ancient wisdom, Ataturk permitted the military institution to decline in relation to civilian power and prestige. Instead, he made the treasury his ally, or at least his preoccupation. As early as 1923 he is quoted as declaring that the great victory won by Turkish arms was not enough to achieve real liberation; the victory had only provided a strong foundation for future victories, victories in science and economics.[8] Again and again he reiterated this thesis.

The maintenance of civilian supremacy in republican Turkey has been a historical fact of the first order. In the Arab lands of the Middle East, as throughout the underdeveloped world, military takeover has been a regular feature of the postwar scene. Turkey's nearest non-Arab neighbors—Greece in the west, Iran in the east—passed through their most severe postwar crises by means of the military coup d'etat. Only in the Turkish Republic had there been no military coup, nor indeed any effective military challenge to civilian supremacy. The coup of May 1960 was the first significant break with the Ataturk tradition.

Apparently on only two occasions had there been military plots directed against the civilian government of republican Turkey. Regarding both occasions the picture is very unclear. First there was the 1926 plot against Ataturk's life, a plot in which at least some military officers were probably implicated. Several high-ranking officers were arrested, including the two generals who had refused in 1924 to resign from the National Assembly as Kemal had demanded and instead chose to resign from their military duties. Both generals were released shortly after their arrest but not before their political ruin. Was this merely a clever attempt to discredit these politically ambitious officers still further? A third officer, a colonel who had been a military aide-de-camp to Ataturk, went to the gallows. Not until 1957 was there again any hint of a military plot against the civilian government —in this case, an alleged attempt to establish a secret organization within the army to incite revolt against the regime. Nine army officers were arrested—three colonels, three majors, and three captains. All, however, were either released or acquitted, with the exception of one major, who was given a two-year prison sentence. Quite obviously, the authorities did not take the matter very seriously.[9]

To explain this unique separation of the political and military requires a comprehensive view of the process which has governed civil-military relations in the Turkish Republic. Although cast in declarative form, our statements have the character of hypotheses. Civilian supremacy had been maintained in Turkey because the governing code of the Turkish Republic was founded upon the mystique of civilian supremacy (which gained the force of a tradition under two civilian presidents who were also the acknowledged military leaders). That mystique had been tolerably well sustained by its visible results; three generations of Turks had acquired values and institutions that flourished better under civilian than military control. In short, by contrast with many other modernizing societies, the military sector in Turkey had been contained because the civilian sector had been relatively successful and the military had nothing better to offer. Since Ataturk and Inonu were the ranking military figures of their day, military interests were well represented at the highest administrative and political level. At the same time, both men were convinced personally of the necessity of civilian supremacy, and they acted consistently on that conviction.

Our evaluation of Turkish developments is, we realize, a direct challenge to those observers who have, in recent years, pointed out the serious deficiencies of the republican regime in Turkey. These deficiencies were real, and the complaints were justified, by American standards or by performance standards based on optimum rationality. The Turks had obviously not attained American standards of performance. Nor had they achieved optimum rationality in policy and practice. *Who has?*

The question is neither impertinent nor irrelevant. Another, perhaps more appropriate, standard by which to judge the Turkish performance is the distance traveled from the starting level toward the level of national aspiration. Viewed in the context of the underdeveloped world today, and particularly by contrast with its Middle Eastern neighbors, Turkey's performance had been relatively good—perhaps even absolutely good. Despite obvious problems, Turkey had met the complex requirements for speeding the tempo of social change while keeping the balance of social equilibrium—that is, for promoting individual mobility while maintaining social stability. How long Turkey can now continue to do so—under any regime—is another matter, particularly in the face of speeding urbanization. But let us return to a more direct consideration of the problem of civil-military relations. We note simply the relative success of civilian leadership in controlling, *while utilizing,* military personnel and resources.

Traditional society in Turkey had been, prior to the Republic, governed by a long-established coalition between the main social orders which constituted the elite—i.e., the court, the clergy, the military. The passing of traditional society began with the crumbling of the court—both the monarchy and its supporting "bureaucracy," the upper echelons of which consisted of an educated class liberated by its landowning wealth for full-time participation in affairs of state. As the court crumbled, the contest for power passed among rivalrous "men of the pen." The learned clergy were a failing force, though in a theocracy they provided the cloak of legitimacy for government and could do much to alienate popular loyalty from any regime not to their liking. While they maintained a strong hold over older people in the rural hinterland, they failed to make effective contact with the newer, younger intelligentsia at the urban centers, many of whom were (or became) young officers infected by Western technology and concepts. As the latter group was relatively small until the end of the nineteenth century, the exercise of power passed, largely by default, to the "civil service"—the administrative institution which often takes transitory control over the affairs of state in periods of decay.

Locked in contest was the "secular intelligentsia," which had not prospered nearly so well under the traditional regime. These included the "free professions"—composed of teachers, students, journalists, engineers, doctors, and the full array of quasi-intellectual roles of a predominantly secular nature. These competing segments of the intelligentsia were engaged in a combat without issue—neither side being capable of sustained governance. The "civil service" was typically too limited by its narrow administrative perspective; the "free professions" were typically too ineffectual by reason of their visionary ideologizing. Neither could claim ultimate authority in a state in which the governing power rested upon religious justification.

When the court crumbles and the intelligentsia is incapacitated, only the "men of the sword" remain capable of taking over political control. For many years in Ottoman Turkey the military had been inactivated by corruption and by division between the traditionalists and the modernists within its leadership. When these problems were resolved, the "men of the sword" moved into political control. W. Z. Laqueur, contrasting the numerous instances of recent military takeover in the Arab countries with that guided by Ataturk, indicates how short the military tether may be: "Military Leadership [referring to Egypt and Syria] thus meant a step forward—but only one and no more. Having forced the old rulers to resign, the new government proved incapable of giving the country the lead which Kemal Ataturk had given Turkey in the early 1920s. The perpetuation of the military rule in the absence of such a lead was tantamount to stagnation, to preoccupation with foreign policy. Absolute rule began to show its corrosive effect on the new rulers, and ever-recurring purges of the officer corps, including the original group of conspirators, caused a rift with the army. It is a process frequently described in the annals of history."[10]

What is the case for Turkish exceptionalism? Why did the Turkish military, having acquired power in the Republic, abjure the role of a military regime? Why had the Turkish military, since then, not been tempted into the alternate course of an alliance between military power and the secular intelligentsia (the modern "revolutionary" version of the classic alliance between men of the sword and men of the pen)? There was, in Turkey, a tradition of such efforts. Writing of the Young Turk Revolution of 1908, L. V. Thomas tells us that "When younger officers of this type decided to join forces with Westernizing true Ottoman civilian intellectuals and with certain 'minority' (millet) leaders of a like mind, Abdul-Hamid's day was done."[11] But the Young Turk Revolution was short-lived. Once the ruling institution was deposed, the incoherent ideology of this coalition and its shallow programing of social change, quickly became apparent. The military were not trained to think in terms of practical application of their high-flown ideas. With power came responsibility—and the coalition collapsed in the debacle of World War I.

Ataturk formed no coalition with the intelligentsia. In any case, they came to his side, as Thomas points out, *after* the seizure of power. Having become boss without their support, he was subsequently free to boss them without the heavy freight of "campaign promises." He used them as he needed them—mainly in support of his own powerful thesis that a modern, participant society operates by the steady enlargement of the margins of reward and responsibility among its citizenry. This is Ataturk's claim to genius as the first planner of social-economic development in contemporary terms. The corollary had to be, if Ataturk's vision was to remain effective, a society based on civilian supremacy in a secular state. In Ataturk's vision both the military and the intelligentsia had a role to play, but their role was subordinate to his twin concepts of populism and etatism. Central political control was essential to regulate institutional innovation—so that changes would occur fast enough, but not so fast that social mobility would undermine political stability. The Kemalist polity was pragmatic. Each policy and program was evaluated by what the ruling elite conceived to be the public welfare, not according to some a priori religious doctrine or political ideology.

How did the army fit into this picture of the Turkish Republic? In the first instance, the army was the vehicle that had brought Ataturk to power.[12] We recall that twelve of the seventeen Ottoman Turkish commanders of World War I became committed to the cause of the Turkish nationalists, and only two remained with the Sultan until the end of his government. But, curiously, Ataturk did not use the army as a major innovating force, after achieving personal power, despite the historic role of the Ottoman army in this direction.

Under the Ottoman Empire, as the vigor of the Ruling Institution waned, the military corps had taken a large hand in modernization of knowledge and institutional innovations. A case in point was the history of the Turkish printing press. Whereas the minority millets (self-governing non-Moslem communities)—Greek, Armenian, Jewish—had introduced their own printing presses into Turkey as early as the fifteenth and sixteenth centuries, a proper Turkish press was not installed until 1728. During the period of continuous and disastrous wars which followed, this press lapsed into disuse. Not until 1784 did Selim III, searching for ways to rehabilitate his declining domains, re-establish a Turkish press. It was used not for spreading popular enlightenment, as among the millets, but for stimulating new thoughts and new ways among the enfeebled elites. Creasy tells us that the revived press specialized in treatises of science and technology relevant for the military corps: "Many European works on tactics and fortifications were translated from the French and published by the Sultan's orders, under the inspection of the Turkish mathematician, Abdurrahim Effendi."[13] Prior to 1909, even Ataturk translated some military treatises from the French.[14]

When Ataturk came to power in 1923, he must have realized, as a military man, that the military victory he had won was only a temporary reprieve. Turkey occupied a piece of real estate coveted by the great powers. To fill the Turkish power vacuum, and to do so before the great powers once again moved against Turkey, would require a total revolution in a relatively short period of time. A modern military establishment was clearly impossible without modern science and industry. These, in turn, were clearly impossible without a transformed political and social system which would permit the people of Turkey to realize more fully their human potential. Illiteracy, debilitating disease, religious dogma and fatalism, subordination of women—all of these had to go. And so the process of dramatic reform in all sectors of Turkish life was set in motion, each individual move being carefully timed and designed to take full advantage of favorable turns of events. Insight, inscrutability, and a sense of timing were among Ataturk's great assets as a radical planner.

Meanwhile, Ataturk carefully isolated the military command from direct involvement in partisan politics. It was not mere happenstance that of the thirty-five men who served in the Nationalist cabinets from 1920 to 1923, only nine were professional military officers, either active or retired (excluding those who had separated from the services and taken up civilian careers).[15] Even that part of the military forces used for rural policing purposes, the Gendarmerie, was placed under direct civilian control within the Ministry of Interior. Officers and personnel of this special corps, although still wearing the Turkish army uniform, were under general staff command only for routine administrative purposes. Within the municipalities, law and order were maintained by a civilian police force which had no direct relationship with the army and was commanded by a separate branch of the Ministry of Interior. Later, a general law was enacted which forbade all private societies based on religion, race, family, region, and social or economic class. Undoubtedly, a military political organization would have been illegal under the general terms of this legislation, although this type of association was not specifically named.

Realizing that the independence and national power of Turkey were based primarily on a sustained social, political, and economic evolution, and that the country was not immediately threatened from without, the leadership diverted relatively little of its total effort to a revamping and modernizing of the military forces during this early period. That portion of the general budget allocated to national defense fell from roughly 40 per cent in 1926 to about 28 per cent in the early 1930s. It has been estimated that as late as 1932 the size of the Turkish army was little greater than that existing in 1922 (78,000 men). It was not until 1939 and 1940 that the Turks mobilized a substantially greater force, possibly something in the neighborhood of 800,000. Also in 1939, when the threat of World War II became a reality, military budgets were again stepped up. The national defense portion then rose to 46 per cent and, in 1940, to 56 per cent, at which high level it remained during the war years. Following the war in 1946, the percentage was cut back to about 33 per cent.[16] But by this time the promise of large-scale American military assistance was very much in the air. Throughout the period, however, the army rested upon nonprofessional soldiers drafted for a limited time from civilian life. Only the officers and some noncommissioned officers were professional.

Although isolated from partisan politics and generally de-emphasized, the army remained an important force in at least two respects. First, economic development plans, especially during the 1930s, were shaped in part by military factors, sometimes at the expense of maximum economic return. Second, the military forces constituted a source of skills and facilities. In times of emergency, military commanders occasionally took over civilian administrative functions—notably in the Kurdish area after the uprisings in 1925 and 1930, and in Istanbul during World War II and the anti-Greek riot of 1955.

Accordingly, many retired army officers were elected to the Assembly or named to high administrative posts during these years. But, as D. A. Rustow points out, "the proportion of them in public life has diminished steadily since the War of Independence." About one-sixth of the Assembly in 1920 was of military origin, about one-eighth in 1943, one-ninth in 1946, one-twentieth in 1950 and 1954, one-twenty-fifth in 1958.[17] On Rustow's calculations, from the founding of the Republic up to 1955, "men of military background have served as President of the Republic (17 years[18]), Prime Minister (16½ years), Minister of Defense (16½ years), of Public Works (15 years), of

Communications (9 years), and in other cabinet posts." Rustow observes that the first cabinet including no officers was that of Hasan Saka in 1948.[19] In addition, many of the provincial governors—the *valis*—were men of professional military origin. In other words, the civilian government did not hesitate to tap the administrative talents developed by the military, but only after the individuals concerned had divested themselves of their military identity.

Nonetheless, as the first Cold War decade opened in 1948, the innovating function which had characterized the nineteenth-century Ottoman officer corps had all but disappeared from the Turkish military. This change was a product of (1) an absence of serious external challenge to the army, (2) the deliberate isolation of the military organization from political activity, (3) successful innovation by the civilian elements, (4) a deliberate de-emphasis of military development by the civilian leadership, and (5) the growth of a military elite based on seniority rather than demonstrated ability. In 1948, the army was still horse-drawn, equipped with World War I weapons, ill-trained, poorly fed, and inadequately clothed. It was no longer the avant garde of modernization as it had been in Ottoman times. Nor was it now a channel of upward mobility around a frozen civilian elite. Civilians were clearly in a position of unchallenged supremacy, politically and technologically, and it was in the civilian hierarchy that ability was most clearly recognized as a basis for promotion and power. Ataturk deliberately appealed to younger men, thereby building an identification between the careers of the more dynamic younger elements and his civilian administration. The military hierarchy froze and went into suspended animation; the civilian hierarchy opened up and led the way to innovation.

Indeed, by 1948, the Turkish military forces were so steeped in tradition that any change was difficult to introduce without reorganizing and remanning the officer corps. Organized along rigid, Prussian-like caste lines, the forces had no personnel policy; rarely were men assigned to tasks on the basis of ability. All high-school and college graduates, other than members of certain ethnic minority groups, could become officers without difficulty, but enlisted men, regardless of proven capacity, could not. Many thousands of soldiers were used as family servants by the officer corps—one instance of the vast amount of military manpower that was wasted.[20] The young Ataturks of the 1923–1948 period would probably not have been attracted to this army.

Relatively low intellectual quality is often associated with a lack of social mobility. A social structure with low performance standards can afford to base recruitment and promotion on status rather than merit. As performance standards are raised, traditional social barriers break down and the key functions go to the men who can do them best, regardless of their social stigmata. In this sense, the American aid program provided a stimulus needed for the revival of the Turkish military corps. With American military aid came a new mission for the Turkish army, and the new concepts and methods needed to fulfill this mission. The impact of this aid on the operational level was expressed by an American military adviser in 1949: "The difficulty is that the men are conscriptees, and the most complicated piece of machinery they may have seen before coming into the service is a wooden stick plough. They hardly know the difference between a hammer and a screwdriver. . . . The Turkish Army will have a tough time keeping the equipment going which we have given it."[21] A further difficulty was the opposition by some older officers to the reorganization and retraining of the military establishment. These officers may have been acquainted with modern military machines from a distance, but few were competent in their use. Hence, modern technical training implied a threat to their leadership.

With the institution in 1948 of American military aid designed to implement the Truman Doctrine, the Turkish army found itself in the awkward position of having its generals receive instruction from American noncommissioned officers, who were, on the whole, as well or better paid and privileged. Halford Hoskins has given us a concise summary of the sequence of events set in motion at the beginning of 1948 by the delivery in Turkey of four large, modern submarines: "Some 350 Turkish naval officers and men were dispatched to the United States for technical training to enable them to take delivery of these and still other naval vessels. This was the commencement of a comprehensive program aimed at the training, both in the United States and in centers established in Turkey, of thousands of Turkish officers and men in the latest American methods of land, sea, and air warfare. The program was the more essential since an estimated 80 percent of the Turkish armed forces was composed of farm boys having no knowledge of mechanized equipment. From this time forward, shipments of military equipment, including training planes, attack bombers, tanks, heavy guns, and the many varieties of material necessary for their maintenance and operation, were scheduled to reach Turkey in a continuous stream, and in such quantities as could be employed effectively by the personnel that had undergone technical training. Although Turkish officials now and then complained that the supply of war equipment fell considerably short of the need, and presently cited this as an argument for the inclusion of Turkey in the Atlantic Pact, American military experts believe that the rate of supply at least equaled the Turkish rate of absorption."[22]

The main mission of the Turkish army was, of course, to provide an efficient force for national defense. To do this, however, as is suggested by the statement just quoted, the Turkish army had to undertake the modernization of the most substantial and significant sector of the Turkish population—its young, healthy males from the rural hinterlands, 700,000 of them. These men had to be taught the difference between a hammer and a screwdriver. They had to be taught how to use equipment of a type which they had never seen or heard of before. They had to learn how to participate in a modern military formation. Fortunately, this new mission coincided with a change in the Turkish political regime (resulting from the 1950 general election), which made possible a shake-up in the general staff and the consequent elimination of some of the older and more conservative officers. The new regime owed no personal debts to the older senior officers. Turkey's participation in the Korean War also added to the general pressure for reorganization, for new types of training—in short, for modernization.

By 1952, the Joint American Military Mission for Aid to Turkey (JAMMAT) had constituted an array of "field training teams," one for each of the three Turkish army headquarters and for each aid-supported division, brigade, and regiment. Such teams were essentially advisory units, the word "inspection" never being used. The need for these teams was revealed by the fact that from 40 to 50 per cent of all vehicles supplied by American aid to Turkey were out of order. The reason was bad maintenance. ("Out of order" in this case did not mean merely "dead-lined" in the American sense; it often meant literally that the wheels would not turn.) By mid-1952, perhaps 90 per cent of the vehicles were again in operating condition.

How was the situation redressed? In part, by the direct intervention of the American field teams. Standard operating procedure required that if a team found something amiss in a Turkish unit—in respect either to its equipment or to its training—the unit commanding officer was "advised" on the spot what should be done. If he refused to conform, the officer was then told that the matter would be reported to Ankara. If he remained adamant, a report went to the Military Aid Mission headquarters in Ankara, where the matter was taken up with the designated members of the Turkish general staff. At the time, one American field team member declared: "Then we almost invariably get action." Turkish unit commanders knew this to be true and, as a result, cooperation on the unit level was reported to be excellent. By 1952 the change in political regime had made possible the designation of younger, more progressive officers who responded positively to efforts to conserve the machines being supplied Turkey under the aid program, even if it meant the active intervention of foreigners on the operational level. This maintenance crisis, plus the support of the Turkish general staff, gave considerable impetus to the Turkish military training program generally.[23] The need for vastly improved technical training had been made dramatically clear.

Out of this process emerged widespread recognition that Turkey's primary need was the rapid and steady increase of its "capacity for absorption"—this involving nothing less than the reshaping of the Turkish people and their perspectives on the world. The thousands of young Turkish farm lads who went through the military technical training programs acquired more than a skill in the maintenance and operation of modern machinery. Many also acquired some degree of literacy and, along with it, the widening of outlook that comes with the acquisition of the three Rs. They acquired new habits of dress, of cleanliness, of teamwork. In the most profound sense, they acquired a new personality.[24] Along with the physical and social mobility opened to them through the military training program, they acquired also the habits of *psychic mobility*. The military corps became, in this decade, a major agency of social change precisely because it spread among this key sector of the population a new sense of identity—and new skills and concepts as well as new machines. Young Turks from isolated villages now suddenly felt themselves to be part of the larger society. The connection between their private lives and public roles became vivid to them —and this sense of their new personality they diffused around them when they returned to their villages. In 1951, anthropologist Paul Stirling reported as follows: "The villagers today are conscious of their membership in the Turkish nation. . . . Under the Ottoman empire, the villagers thought not in terms of a nation, with a national territory, but of themselves as a group distinct from their neighbors on grounds of language or religion, or both. Whatever its origin, the present villagers are decidedly nationalistic; proud of Turkey's military prowess, interested in her mechanical achievements. . . . Undoubtedly, consciousness of being part of the whole nation is greatly strengthened by the compulsory term of military service which all the young men seem vastly to enjoy."[25]

By 1952, one could list eleven major changes that had been

accomplished or at least incorporated into the thinking of the Turkish general staff: (1) reduction of the Turkish army from an estimated 700,000 to 400,000; (2) inauguration of service training schools and great expansion of technical training generally;[26] (3) the training abroad of a significant number of officers and noncommissioned officers, many of whom would later become instructors;[27] (4) institution of a testing and classification system as a basis for assignments; (5) organization of a personnel section in the general staff and assignment of separate personnel officers to each major unit headquarters; (6) reorganization of the Reserve Officers' Training School; (7) requirement of a year's troop duty before admission to the Military Academy; (8) reduction of the initial stint for noncommissioned officers; (9) a new noncommissioned officers' law, making possible officer status for qualified NCOs; (10) elimination of the use of soldiers as personal servants by officers; and (11) much-improved living conditions for enlisted men.[28]

The first three items indicated a major and realistic reconsideration of Turkey's "capacity to absorb." Cutting the size of the army almost in half, while simultaneously improving and expanding technical training, bespoke a determination to do better by each man who passed through military training—a move toward making him a qualified modern soldier and, upon his return to civilian life, a qualified modern man. The third item likewise implied an accelerated transmission throughout Turkey not only of Western military and technical skills but also of modern political and social concepts. These would be passed on to the ordinary soldier during the course of his training. The next four items were clearly designed to make superior training reach its payoff points by adequate methods of personnel selection and job classification. The last four items were important measures of democratization. While these changes were occurring, a couple of million men passed through their military training and a comparable number will do so during the next decade.

Consider the impact upon these young men of their first exposure to the wider world outside their village, as they undergo training in this rapidly modernizing and democratizing army. Granted, as the result of a decade of both military and civilian training, it is no longer true that 80 percent of the army's recruits are farm boys with no knowledge whatsoever of modern machinery. But the multiplier effect continues to operate, since 50 percent or more of the recruits are still rural illiterates gaining their first exposure to a social institution operating in the modern style. Not only is this an army in which enlisted men might become officers (although this still happens rarely in practice), but even the members of ethnic minorities are now accepted as candidates for officer status. In the fall of 1955, indeed, several *women* were admitted to Turkey's War Academy as cadets.

All of this new egalitarianism takes place within the context of what is to illiterate and impoverished boys from the villages an overpowering "materialism." One has the impression that, owing to a projection of our own Western doubts and dilemmas, the value of such "materialism" for modernizing countries has been misunderstood and maligned. The "revolution of rising expectations" is motivated largely by the ever-spreading desire for more of the good things of life among people who, through most of history, neither had nor expected to have these things.

What happens when a young Turk returns to his village after having served his two years in the army? In the absence of the definitive case studies, one can only make general observations. Clearly, if army living standards are superior to those prevailing in much of the surrounding civilian society—as now seems to be the case—the young discharged soldier is likely to find his traditional society inadequate to sustain his new level of expectations. A sense of frustration stimulates the young man to apply his army-acquired skills to promote civil innovation and "progress." He has been exposed to the machine concept and is aware of the advantages accruing to those with mastery of mechanical skills. Furthermore, the prestige of his military service has become identified in all minds with the new technology. Hence, it is likely that the soldier returning to his native village will resist falling back into the premechanical era.

This resistance may lead to a falling-out between father and son. Contemporary observation in selected Turkish villages indicates that the apparent replacement of extended-family units by primary family households may be partly due to this friction. Young men are breaking away physically from their fathers' households, prior to the death of the father, and setting up their own homes perhaps a few hundred yards away. This practice is a distinct break with tradition. Moreover, many younger men are leaving the villages permanently and seeking a new life in the market town or big city. Recently, in private conversation, a high Turkish government planner speculated that one of the driving forces behind the accelerating urbanization is the desire of the younger people for an independent life in the primary family, an independence difficult to achieve within the small, semi-isolated, traditionally organized rural communities.

If the young man returning from the army—even though breaking away from his father's household—remains within the village, he may be the driving force behind both technical and social innovation, such as the purchase of modern agricultural equipment and the organization of a local youth society. He may also possess the necessary skills to operate the tractor. At the very least he now has the technical background that renders him a likely candidate for a short training course in the operation and maintenance of modern agricultural equipment. In other words, the absorptive capacity of this civilian sector in Turkey has been increased by the military training program.

As the number of such young men grows, the prestige of the military becomes identified increasingly in the public mind with technical competency. This identification can be of prime importance, for the military ranks high in Turkish eyes. On the basis of systematic field work, George Helling concludes that the dominating image for modern Turkish youth is still the soldier.[29] Therefore, identification of technical skills with military competency is perhaps of greater significance in Turkey than it would be in certain other societies.

Even general education may be identified with the military in the public mind, by reason of the large-scale program of basic education now being instituted by the Turkish army. The major purpose of this effort is to submit all illiterate draftees, who demonstrate in tests the native intelligence to participate successfully, to a seven-week curriculum in reading, writing, arithmetic, measurement, and elementary social studies. The program is just now getting into full swing, which means that roughly 120,000 men (50 percent of the draftees) will go through the course each year in sixteen specially constructed, army education centers distributed over the country. The present course, given to draftees immediately after induction, includes thirty hours of Turkish per week, three hours of social studies, eleven hours of arithmetic. It is predicted that roughly 70 percent of the students will complete the course successfully. Special textbooks have been published with the cooperation of an American specialist. Later, over 100 additional titles will appear in simple Turkish for distribution to soldier-graduates to encourage continued reading.[30] Student motivation is quite obvious. One journalist reported his visit to the first pilot class in April 1959 as follows:

> Visitors to a classroom were welcomed by this rehearsed teacher-pupil exchange in Turkish:
>
> "What will you be if you do not learn to read and write?" asked the teacher.
>
> "Privates," chorused the class.
>
> "And what will you become if you study hard and learn your lessons?" pursued the teacher.
>
> "Corporals and sergeants," the class answered enthusiastically.[31]

There is reason to expect that this army program may give general education in the villages and small towns a very healthy shot in the arm by raising public acceptance, and even demand, for secular education. This will represent a vast change in attitude from the early days of the Republic, when secular education was resisted vigorously by large elements of the general public. Even today, there is considerable passive resistance, often religiously inspired, to the concept of secular education in the more remote rural areas, particularly in the eastern provinces.[32]

By 1959, approximately two billion dollars' worth of American and European equipment and technical assistance had gone into the shoring-up of Turkish military strength. Meanwhile, increasing attention had been directed to building morale within the army and a more careful consideration given to the total impact the army might have on economic and social development. In a long statement in February 1958,[33] the Turkish Minister of National Defense explained that the new law on personnel assignment policy was being deliberately applied in such manner as to broaden the geographical experience of army personnel. In essence, the country was divided into three districts—western, central, and eastern. Within each district, military posts were classified into four categories depending upon their general environment—economic and social, climatic, cultural, and medical. Before an officer could be assigned a second time to the first district, he must serve out an assignment in either the second or third. Likewise, the classification of his specific post must change from assignment to assignment. (For some time, a general policy had been followed to the effect that draftees from the cities should be assigned to rural areas, rural draftees to duty in the cities.[34]) The Minister also reported improvement in military housing and health conditions. He made significant reference to army and navy efforts to produce in military factories certain machine and vehicular parts required by the civilian economy, "in order to make possible the participation of the military industrial establishments in our general development."

One should not discuss the impact of the army on the total economic and social structure of the country without reference to the highway development program, initially considered almost exclusively as a military matter. The U.S. Highway Mission dispatched to Turkey in 1948 was placed first under the Military Aid Mission. Not until sometime later was the highway program related to the economic aid mission. Certainly among the reasons that Turkish leadership became so interested in pushing the highway program were its military implications. Highway development has now gone far beyond that phase and is justified largely in economic and social terms. It is possibly now true that the vast majority of the Turkish population resides within relatively short distances of all-weather roads. Obviously, the existence of these roads has had an enormous impact on the routine of life in the rural communities, and hence also in the market towns and cities. Many hundreds of thousands of villagers now travel into market towns quite frequently simply to pass the time of day, to sit and chat in a coffee house, or to attend a local cinema. This increased geographical mobility is not unrelated to increased social and psychic mobility.[35] The new technical training in the Turkish army has provided many thousands of drivers and mechanics for the new vehicles moving on these new roads. This development of increased opportunity for geographical mobility has, of course, strengthened the tendency that we have already noted toward a breakup of the extended-family organization within the traditional community. It might also be observed that specialization in military function has paralleled and, to a certain extent, induced a similar occupational specialization within the civilian sector.[36]

An outgrowth of the use of the military to stimulate general development has been the Turkish emphasis on the nonmilitary aspect of what are fundamentally international military pacts. As originally conceived, the Balkan Pact was to have been implemented by a number of nonmilitary commissions active in the economic and cultural and political fields—one purpose being to expand trade relations among Greece, Yugoslavia, and Turkey and hence stimulate economic development generally. Also, in connection with the Baghdad Pact, now called the Central Treaty Organization (or CENTO), the Turks have been instrumental in directing attention to nonmilitary activities, specifically the construction of an adequate communication system between Iran and Turkey. This project could be of substantial importance to the Turkish economy by opening up the easternmost provinces of the country, wherein are found rich mineral resources.

The Turks have also used the army as an instrument for general cultural development in the immediate areas surrounding major army installations. The location of the First Army headquarters in Erzurum has stimulated the effort to develop that city as the cultural center of eastern Turkey. Among the specific reasons cited by Turkish authorities for the establishment of a new university in Erzurum was the existence there of large numbers of military children, who would be likely candidates for university education. The impact of such an educational enterprise on eastern Turkey may be considerable. In recent years, the growth elements in these areas—the more educated and more wealthy—had been lured increasingly into Ankara and Istanbul by the more comfortable life and greater intellectual companionship to be found in these cities. Those parts of the country most in need of their dynamic elements were thus deprived of them. The army now tends to reverse this flight from the rural east and to provide a new dynamism that is likely to speed modernization by using public resources to build institutions that are educative, egalitarian, and functional.

Perhaps there is in the Turkish experience something of general import for American policy in economic development. One suspects that the distinction between economic aid and military assistance can easily be exaggerated—and is, in any case, ambiguous. We can deliberately plan, and help others to plan, our military assistance for maximum effect upon the economic and social development of the recipient country. For example, foreign military personnel coming to this country for professional or technical training should perhaps also be exposed to our economic processes, political institutions, and social practices. Indeed, the army, the public school, the industrial enterprise, and the economic planning organization of an underdeveloped country tend to converge on common objectives. In such a situation, professional military training may well come to differ somewhat from that which is most appropriate in the United States or Western Europe. Perhaps, learning from the Turkish experience, we can help other countries to harness their national motivation to be strong, via military organization, more directly to the drive for accelerated development.

Finally, recent events raise questions about the fundamental compatibility between a vigorous military institution, on the Turkish model, and the principle of civilian supremacy in government. Why, after such a splendid history of service under civilian authority, did the Turkish army take over power in May 1960?[37] Does it now intend to install long-term military rule or will it, as it claims, create anew the conditions under which a democratic polity and civilian supremacy can function efficiently? Any answer to these questions involves an appreciation of the historic phase through which Turkish society, and its military institution, have passed.

The Turkish army has been nourished by, and in turn has nourished, that type of modern materialism which places a high value upon personal well-being. This "revolution of rising expectations," which has been spreading rapidly around the world, has turned out to be extremely treacherous in many areas. Expectations are raised long before they can be satisifed. The result is, quite often, a "revolution of rising frustrations." In one underdeveloped area after another, the inadequately satisfied or inhibited aspirations of the new literates have cumulated into a persistent discontent that makes stable parliamentary government under civilian political leadership impossible. This has led, in one country after another, to military takeover and the imposition of stability by coercion.

Until May 1960, the Turkish case was exceptional. The army, while adding its bit to the rising tide of expectations, had helped provide the perspectives, the skills, and the roles which could satisfy these expectations. In so doing, it had moved farther and faster than the army (or the civilian sector) had been able to do in many other underdeveloped countries. The evidence seemed clear that the Turks had used the army successfully as a lever for general development while maintaining republican institutions and civilian supremacy.

The Turkish army seemingly had, almost uniquely, abandoned the military prerogative of standing aloof and watching civilian institutions crumble, and then taking over after the civilian elites have been brought to a complete impasse. It had, rather, committed itself to the firm and efficient support of a rapidly modernizing and democratizing society under the rule of civilian supremacy. This was exemplified by its commitment to recruit, train, assign, promote, and release the 400,000 men under its charge on the democratic basis of merit and the efficient basis of ability. Such commitment, in any military corps, acknowledges civilian supremacy in defining the goals and regulating the larger processes of the society as a whole.

Why, then, did a military junta proceed to wrest political power away from a civilian government during the predawn hours of May 27, 1960?

The top priority of Turkish policy, in the past several years, has been to maintain a balance between "level of aspirations" and "capacity for absorption." Whatever its short-term defects might be, this is a reasonable policy for any underdeveloped country caught in the rapid tempo of modernization. Once the rising tide of expectations gets out of hand, once people want more than their institutions and available resources can provide, then the special brand of chaos that we have seen in many underdeveloped areas occurs. In this sense, Turkish materialism seemed a very promising avenue to a stable future based on the well-being of its citizenry.

The Turkish elite, in all the key sectors, had shown a profound awareness and consensus on this point. This awareness underlay what we consider to be the most remarkable demonstration of successful party politics in any underdeveloped area since World War II, despite the military coup of last May. Whatever may be said, and justly, in complaint against the harshness of the Democratic Party's policies toward the press and civil liberties, it had not permitted itself to deviate from the course of modernization based on economic development. Whatever might have been the anxieties of the People's Party to regain office—the most natural desire of all "out" parties being that they want to be "in"—its leadership had not succumbed to the politically attractive, if nationally dangerous, temptation to make political hay of traditionalist religious and xenophobic sentiment in the rural hinterlands. Nor had the Democratic Party, until recently, attempted to use the military to support partisan power—despite the uncertain outcome of recent elections.

These restraints can be taken as evidence that the Turks and their leaders, perhaps more than most others in the vast underdeveloped world, had defined what it was that they really wanted. Official Turkish spokesmen had stated privately that if forced to choose between continued economic or military development at the present pace, they would choose the former. They felt that their first line of defense, both against becoming too dependent on the West and against possible loss of sovereignty to the USSR, lay in continued economic growth at a pace that would bring Turkey, within the foreseeable future, into the range of the Western industrial states in terms of productivity and standard of living. These spokesmen had expressed the fear that in the absence of such economic growth, the days of Turkish independence would be limited. Indeed, economic development seemed to have been elevated to the position of a national ideology, to which the army subscribed as enthusiastically as did other elements in the population.

The lesson to be derived from the current failure of civilian leadership in Turkey is that such leadership can survive only as long as it continues to lead. The Turkish army could be entrusted with major programs that shaped national development—basic education, technical training, industrial production—so long as the civilian leadership generated the ideas and shaped the institutions which made these service programs functional for the society as a whole. The government made policy; the army served it. So it was for thirty-seven years. But when civilian leadership seemed to falter—(1) by leaning upon the army for political support, (2) by clogging channels of promotion for young men of ability and subjecting them to partisan pressures, and (3) by alienating the intellectual elite, which was closely identified with the junior officer corps—when these things happened, the situation changed. Moreover, since 1954 the civilian administration had been losing its former appeal to the growth elements in the population—the modern intellectuals, the ambitious technicians, the innovators and entrepreneurs. As the civilian administration was becoming a closed society (with clique values more important than ability and achievement in pushing men ahead), opportunities in the military establishment were increasingly becoming available to men of ability regardless of their origin, family, party, or age. The army's dynamism was not matched by the civilian sector. This tended to make the army's programs dysfunctional; its satisfactions turned into frustrations. When the tightly closed regime then tried to use the army for its own partisan political purpose—thereby violating the basic Kemalist doctrine of an apolitical army—it subverted the principle of civilian supremacy. The only way to save civilian supremacy under these conditions was, paradoxically, a military coup to install a caretaker regime that would re-establish the conditions of democratic civil government. What are the prospects?

The new military regime in Ankara claims to be reform-minded, not vindictive or partisan. The Committee of National Unity, the thirty-eight-member military group responsible for the seizure of power and administration of the provisional government, has uttered unequivocal commitments to republican principles and has promised to dissolve itself on the day that a new parliament convenes. By midsummer it was known that a number of important constitutional changes were to be undertaken—namely, addition of an upper chamber of the Turkish parliament, formation of a court empowered to review the constitutionality of legislative acts, institution of a proportional system of representation. Also, political prisoners had been released and the press had been freed (although not entirely so). On the other hand, virtually the entire leadership group (403 persons) of the Democratic Party had been imprisoned.

Despite the idealistic atmosphere of the Turkish coup, we know that power corrupts. In speculating on the Turkish future, one must take account of certain elements in the prehistory of the coup. In the first place, there is no record that General

Cemal Gursel, present dictator of Turkey, resisted the use of the army for political purposes by the previous Inonu regime (1938–1950), nor its corruption, its rigged elections, and its almost total disregard of constitutional rights. The General apparently became disaffected with the Menderes regime shortly after he had been passed over for the top post in the army. He had been in military service for forty-five years, a general since 1947. Moreover, the opposition Republican People's Party had maintained an unremitting campaign at pre-election intensity against the government since the 1957 election, in which the party had drawn 38 percent of the popular vote. Every act of the Menderes government—other than foreign policy—was condemned. The Menderes regime reacted by forbidding public political meetings except during a pre-election campaign. Ismet Inonu's deliberate flouting of the law in Kayseri on April 3 instigated the use of the army to prevent his further campaigning. (No election was in the offing.)

On March 21, the government reportedly (not refuted) intercepted a secret order from the headquarters of the Republican People's Party to its branches virtually directing them to go underground, to name secret couriers, and to found an "ear newspaper." On April 17, Ismet Inonu told a group of retired generals and admirals (in his home) that it was up to them, and to the military, to protect the soundness of Turkish society and the ideals of Turkish progress and development. This statement was reported in the press and has not been refuted. Two days later, government spokesmen understandably charged the opposition with deliberate conspiracy to overthrow the government by force. It was the appointment of a commission to investigate these charges and the suspension of all partisan political activity for ninety days that touched off the student riots. Republican members of parliament were found among the students during the rioting, thereby casting doubt on the spontaneity of at least some of the disorders. Charges made about murdered students during the time of the riots were later found to be unsubstantiated. (As Adnan Menderes publicly announced at the time, two were killed during the rioting, almost accidentally.)

In his letter of May 3, which was virtually an ultimatum, General Gursel insisted that President Celal Bayar should resign, but no mention was made of Prime Minister Menderes. He also insisted that "problems of concern to the Army must be solved with dispatch." The use of the army for apparent political purposes was mentioned elsewhere in the document. One wonders what these other, undesignated problems were. All except three members of the entire parliamentary delegation of the Democratic Party were imprisoned shortly after the military takeover and are to be denied freedom to participate in the new elections, thereby virtually assuring domination by the Republicans—a minority party insofar as is known. However, there is considerable evidence that, right up to the moment of the coup, the Democratic Party remained the popular party. Prime Minister Menderes was attributed on the grass-roots level with having diverted an increased share of the national income out of the cities and into the villages and small towns.[38]

Apart from the subversion of civilian supremacy, the military takeover sets several exceedingly dangerous precedents: (1) the trial of ousted politicians for "unconstitutional" or "illegal" acts, (2) the collective punishment of a regime, and (3) constitutional reform without popular mandate. Should the coup eventuate merely in the transfer of power to the Republican People's Party, the spirit of rancorous reprisal against political opponents, long an element of public life in Turkey, will remain alive. Any Republican administration holding office in the near future will hardly dare relinquish power to a Democratic administration, even though the latter be the victor at the polls. The short-run costs to democratic polity of a coup made in the name of democratic polity will be high. For the long run, much will depend upon the wisdom of the present caretaker regime in setting conditions under which democratic political life can grow rapidly and civilian supremacy can prosper.

The Turkish experience vividly illustrates the constant challenge to civilian leadership posed by a rapidly developing society in which the democratic tradition is new. The striking past accomplishments of the Turkish Republic enhanced, but obviously did not automatically assure, its prospects for creative civilian leadership in the future. The new nations-in-a-hurry seek to accomplish in decades the modernization that was achieved over centuries in the West. Their great need is to accelerate mobility while maintaining stability. The military institution, historically the chosen instrument for imposing stability, can be transformed into an agency that promotes mobility within a framework of civilian ingenuity and innovation. But when the civilian leadership fails to maintain the tempo and balance of social change, when dynamic growth is crippled by partisan nepotism, a frustrated military is likely to take over control. This is the great lesson which the Turkish Republic has to teach the underdeveloped nations of the world. Its past provides a model, and its future a test, for the role of dynamic government in the critical decades of rapid growth that lie ahead—and, perhaps more fundamentally, for the supremacy of civilian authority itself.

NOTES

1. A. H. Lybyer, *The Government of the Ottoman Empire in the Time of Suleiman the Magnificent* (Cambridge, Mass., 1913), pp. 90–91.
2. H. A. R. Gibb and H. L. Bowen, *Islamic Society and the West* (London, 1950), p. 31.
3. Quoted in I. Orga, *Phoenix Ascendant* (London, 1958), p. 38. See also "Ataturk," *Islam Ansiklopedisi* (Istanbul: Ministry of Education Publishing House, 1940), p. 271; and L. V. Thomas, "The Political Leader: Ataturk: A Case in Point?" a paper presented to a conference on "The Near East, Social Dynamics and a Cultural Setting," sponsored by the Committee on the Near and Middle East, Social Science Research Council, at Princeton University, October 1952, p. 5 (typewritten).
4. Orga, *Phoenix,* p. 39.
5. Mustafa Kemal, *Speech of October 1927* (Leipzig, 1929), p. 692.
6. See Law No. 5398, May 30, 1949.
7. Article 8, Election Law of 1946; Article 9, Election Law of 1950.
8. *Ataturk Diyor Ki* ("Ataturk Speaks"), (Istanbul: Varlik Yayinlari, 1951), p. 86.
9. *Cumhuryet* (Istanbul), November 26, 1958.
10. W. Z. Laqueur, *Communism and Nationalism in the Middle East* (New York, 1956), p. 22.

11. L. V. Thomas and R. N. Frye, *The United States and Turkey and Iran* (Cambridge, Mass., 1951), p. 53.
12. D. A. Rustow, "The Army and the Founding of the Turkish Republic," *World Politics* XI, no. 4 (July 1959), 533.
13. E. S. Creasy, *Turkey* (New York, 1906), p. 377.
14. *Ataturk'un Askerligi Dair Eserleri* ("Ataturk's Military Writings"), (Ankara: Is Bankasi, 1959).
15. Rustow, "The Army," p. 535.
16. Figures are taken from *Mali Istatistikler* ("Financial Statistics"), (Ankara: Basbakanlik Istatistik Genel Muduru, 1947), and *Commentary on the 1950 National Budget by the Ministry of Finance,* trans. Economic Cooperation Administration Special Mission to Turkey (Ankara, 1950).
17. Rustow, "The Army," p. 550; see also the *Album* for 1950, 1954, and 1958, published by the Grand National Assembly in Ankara. An article entitled "Military Deputies in the Assembly," in *Halkci* (Ankara), February 3, 1955, gives short biographies of the fifteen retired military men sitting in the Tenth Assembly. Two of these had also pursued civilian careers in law and government administration.
18. Rustow must mean 27—Ataturk from 1923 to 1938, Inonu from 1938 to 1950.
19. *Ibid.*, p. 530.
20. See R. D. Robinson, "American Military Aid Program," New York, Institute of Current World Affairs, November 15, 1948; *idem,* "Impact of American Military and Economic Assistance Programs in Turkey," New York, American Universities Field Staff, 1956. For a graphic account of the relatively low level to which the army had fallen by World War II, see A. Humbaraci, *Middle East Indictment* (London, 1958), pp. 36–40.
21. Robinson, "American Military Aid Program."
22. H. L. Hoskins, *The Middle East* (New York, 1954), p. 37. See also press statement by Major General Horace L. McBride, Joint Military Mission for Aid to Turkey, Ankara, August 18, 1950.
23. R. D. Robinson, "Americans in Turkey," letter to the Institute of Current World Affairs, New York, June 1952.
24. Based on interviews with Lt. Col. Leon D. Marsh, Classification Officer and Assistant G–1, TUSAG, Ankara, during January 1952.
25. P. Stirling, "The Structure of Turkish Peasant Communities" (Ph.D. dissertation, Oxford University, 1951; reproduced in Ankara by R. D. Robinson, mimeographed), p. 165.
26. Shortly after the arrival of the American Military Aid Mission in 1947, a number of technical schools were established or greatly expanded, including schools designed to train technicians in naval electronics, naval ordnance and gunnery, mine warfare and sweeping, naval damage control and fire fighting, naval shipyard and industrial operations, submarines, antiaircraft defense, army signal work, army medicine, army ordnance, infantry, artillery, army motor transport, army engineering, pilot training, radio communications, aeronautical mechanics, meteorology, aeronautical supply, and the operation of heavy highway equipment.

27. No exact figures are available as to the number sent abroad, but as of early 1951 over 1,000 Turkish military personnel had been trained in American methods either in the United States or in Germany. (R. T. Hartman, *Uncle Sam in Turkey* [New York, Turkish Information Office, 1952], p. 14 [a reprinting of articles published in the *Los Angeles Times* during September 1951].) Early in 1958, the Turkish Minister of Foreign Affairs announced that 300 Turkish jet pilots had been trained in Canada alone. (*Zafer* [Ankara], February 26, 1958.)
28. Humbaraci, *Middle East Indictment,* p. 39n, attributes the improvement of conditions in the Turkish army during the 1950s to a short spell of political liberalism and the experience of several thousand soldiers in Korea, where they had seen armies with very different standards, as well as to the influence of the U.S. Military Mission.
29. George Helling, "A Study of Turkish Values," a paper delivered at the Conference on Turkey, Harvard University, June 1959.
30. Interview with Colonel Mithat Ceylan, officer in charge of the program, Ankara, October 2, 1959; see also Emin Hekimgil, "Education in Turkey," a paper prepared for the 22nd International Conference on Public Education, Geneva, July 1959, pp. 33–34.
31. Jay Walz, in *New York Times,* April 17, 1959.
32. K. Y. Wofford, "Report on Rural and Primary Education and Related Teacher Training in Turkey," a report to the Turkish Ministry of Education, Ankara, 1952, p. 30 (mimeographed).
33. Reported in *Zafer,* February 25, 1958.
34. Described in *Cumhuryet,* July 16, 1957.
35. The impact of roads on the spread of Turkish mobility is described in Daniel Lerner, *The Passing of Traditional Society* (Glencoe, Ill., 1958), chaps. 1, 4, 5.
36. R. D. Robinson, "Tractors in the Village," *Journal of Farm Economics* XXXIV, no. 4 (November 1952), 451.
37. Described in R. D. Robinson, "Upheaval in Turkey," *Foreign Policy Bulletin* XXXIX, no. 20 (July 1, 1960), 153 ff.
38. As an apparent step to curtail popular participation in political life, a temporary law went into effect on July 4, 1960, which forbade political parties to set up or maintain branch organizations of any sort other than those located in provisional seats and municipalities. This move was obviously designed to break the organizational strength of the ousted Democrats in Turkey's 38,000 villages, wherein dwell 70 percent of Turkey's population. Preceding the move were recurrent reports of arrests in rural areas occasioned by uprisings and demonstrations in favor of the jailed leaders.

5 *The Army, Politics, and the People*

A. ISKENDEROV

The fact that the army has recently appeared in the foreground of political events in many young national states of Asia and Africa has posed a number of problems whose theoretical elaboration acquires great practical importance in working out the paths, forms, and methods of uniting the progressive forces in the struggle against imperialist and internal reaction, for national and social progress. To this group belong, in particular, questions of the army's role and place in the national liberation movement, its influence on the political life of the developing countries, and importance in the struggle of the popular masses and their organizations.

There are various interpretations of the army's role in the national liberation movement. While some authors refuse to attrib-

From *Izvestia,* January 17, 1967. Translation copyright 1967 by *The Current Digest of the Soviet Press,* published weekly at Ohio State University by the American Association for the Advancement of Slavic Studies; by permission.

ute any progressive role to the army and define political power established through military coups merely as reactionary and militaristic, others attempt to portray the army as almost the only all-national force capable of heading a national liberation revolution. Sometimes it is even said that the army is seemingly in a position to direct the political and economic development of the liberated countries independently on its own, as it were, without the support of the broad masses of working people, without an advanced political party.

Both of these viewpoints are unsound, since their adherents confine themselves to a superficial grasp of events without taking into account the deep social and economic and political processes unfolding in the countries of the so-called third world, evidently forgetting that the nature of any social movement is determined not by who leads it but the *objective* result of its development and the aims it *objectively* serves.

The class approach is imperative in appraising the army's role in the national liberation movement or any other social phenomenon. The army's position in society is determined by the nature of the society itself, the correlation of class and political forces, the level and scope of the revolutionary movement. In any class society the revolutionary struggle of the working masses inevitably draws the army into the whirlpool of political events. As V. I. Lenin stressed, "The army cannot be, never was, and never will be neutral."

The experience of the national liberation movement in Asia and Africa shows that armies are not only not neutral but actively participate in the political life of their countries, often exerting decisive influence on the development of events. A number of factors are conducive to this.

First, the low level of social and economic development of many of the recently liberated states, insufficient class differentiation, and immaturity of social relations brought about a situation unique in that one social class could not direct the revolutionary process single-handedly. The working class could not do so because of its small numbers, insufficient political maturity, and weak organization; the national bourgeoisie could not because it has not yet taken shape as a class at all in some countries, and wherever it did head a national liberation movement, its advent to power was not, as a rule, accompanied by the strengthening of the liberated states' economic and political stability. In these conditions there is an objective growth in the role and significance of the state and those of its weapons, like the army, that acquire relatively greater independence regarding classes than in countries with highly developed class structures. "By way of exception there are, however, some periods," F. Engels wrote, "when the embattled classes attain such an even balance of forces that for a while state power has a certain independence regarding both classes as a seeming mediator between them" (K. Marx and F. Engels, "Works," second edition, vol. XXI, p. 172). This proposition of Engels is applicable not only to societies with developed class structures but, also, and probably even more so, to countries where class differentiations have not emerged with sufficient clarity, particularly to characterize the situation that has evolved in certain liberated countries that are undergoing transitional periods in their historic development.

Second, the army did not stand apart from the struggle of peoples for national liberation. Freedom-loving ideas penetrated its ranks early. In many countries of Asia and Africa the national armies were formed right in the midst of the revolutionary liberation movement and hence stood forth as anti-imperalist and progressive forces from their very inception. This is what happened in Algeria, where the Army of National Liberation, composed of poor peasants, workers, and petty bourgeoisie, waged an anticolonial war, expressing the interests and hopes of the entire Algerian people. The backbone of the army that fought for the independence of Burma, the army that played an outstanding role in freeing that country from the Japanese occupation forces and the British colonialists, was also composed of peasants, workers, and the urban poor.

As for the officers, the most enlightened and active section of the army, many of them were only loosely connected with the class of landowners and big bourgeoisie. They were mainly the offspring of the petty bourgeois milieu of cities and villages, that is those strata of society whose social positions in economically backward countries are much closer to the toiling masses than to the bourgeois elite. This accounts for patriotic officers' hostility not only to imperialism and colonialism but to capitalism as well. The officers were able to go abroad and acquaint themselves with advanced ideas. Returning to their homelands, they felt especially sharply the necessity of political and social and economic changes, even though they still had no clear-cut program for constructing the new society.

Third, during the struggle for national liberation and even more so after winning state independence, the political parties in many liberated countries were unable to exercise decisive influence on the course of developments in their countries. In most cases these parties, which more precisely were organizations of the national front type and united diverse social, political, and ideological forces, were not adequately prepared to solve the incomparably more complex tasks confronting the people who had won state independence.

To a certain extent these factors explain why in many developing countries the army emerged on the crest of political events as the most organized force of society.

At the same time it would be wrong in both theory and practice to lose sight of the fact that the army's functions and position in society cannot help changing with the development of the national liberation movement, when the solution of social and economic tasks advances to the foreground, when questions of economics acquire increasing importance for the young national states. Army leaders who came to power through the army are beginning to understand that the army, whose activity is based mainly on orders and instructions, cannot take the place of the democratic institutions of society, without which national revolution cannot successfully progress. Confronted with reality, many of them grew increasingly convinced that the army, with no clear political or ideological platform, no experience of political organization of the masses or economic construction, and no experience in leading the class struggle, cannot take the place of a party as the guiding force of society, that the construction of a new life and the victorious struggle against imperialism and neocolonialism were impossible without the active participation of the laboring people and their most advanced detachments.

The construction of the new society is a long and intricate process that is the more successful the more broadly it enlists all progressive forces, organized so that all or almost all of the classes and social groups can make their contributions to the country's progressive development. The mixed economies of most of the countries being liberated or already liberated, the complex pattern of different forms of property—from primitive communal to capitalist property—and the great political, tribal, and religious diversity of the populations of these countries make for a prolonged transitional period that could occupy an entire historical epoch charged with acute class conflicts. As the Twenty-third C.P.S.U. Congress emphasized, "The new life in the liberated countries is being born in fierce clashes with imperialism, in sharp struggle between the forces of progress and the forces of reaction within these countries."

Aside from the external factor, namely the expansion and strengthening of the positions of world socialism, primarily its successes in economic competition with the world capitalist system, such an important internal factor as an advanced political party's leadership of the revolutionary movement has decisive importance for substantial reduction of the transition period and acceleration of the process of social development in the liberated countries. A revolutionary movement can arise and develop without an advanced political party. But to consolidate the victory of the revolution and construct the new society, a party is needed that would reflect the ideology of the working classes and would be guided in its activity by a knowledge of the objective laws of social development. The army, which is confronted by specific tasks connected primarily with defending the revolution from encroachment on its achievements by external imperialist forces as well as internal reaction, cannot take the place of such a party. In addition, the army, which is not a unified whole socially and ideologically, falls into two opposed camps as soon as the question of selecting the path for the country's further development is raised, for the process of the demarcation of class forces, intensified after the achievement of political independence, has affected it too.

In this situation one segment of the patriotic officers, reflecting the attitudes of the popular masses and enjoying their support, takes decisively to the path of further development of the revolution and implementation of profound social and economic and political transformations paving the way for the subsequent progress of their countries toward socialism. The other segment of the officers, influenced by imperialist or internal reaction, either play a waiting game or openly oppose the revolution. The lessons of military upheavals in several countries of Asia and Africa have shown that the army can not only play a progressive role in the national liberation movement but can also easily turn into a tool of reactionary forces, if the influence of democratic ideas is weakened. In the one case the army serves to accelerate progressive development in the liberated states, and in the other, on the contrary, to hamper that development. The army is a societal institution in which democratic ideas can live rather placidly alongside reactionary views, thereby breeding the danger that individual holders of these views may attempt to use the army, or a part of it, against the revolution. Which of these tendencies in the army, the democratic or the reactionary, will gain the upper hand at a given historical stage depends on many factors, but primarily on the level and scope of a revolutionary movement capable of engendering ferment within the army and winning the whole army, or a substantial part of it, over to the side of the revolution.

In a number of cases, when a military coup is carried out its leaders proceed from purely opportunist motives of coming to power. It is not surprising that the imperialists follow the development of such tendencies within the army attentively and attempt to use them for their own neocolonialist purposes.

One cannot fail to see, for instance, the connection between the recent events in Ghana and the subversive activities of the imperialist powers in Africa, who count greatly on the army. They make skillful use of the circumstance that the character and composition of the armies of many African countries have changed little since the colonialists left. The armies continue to be mostly mercenary, and the training and instruction of the soldiers are left essentially to Western "instructors." Underestimation of political work in the army and the attitude that it is a force allegedly outside politics can have serious consequences for the destinies of national revolution. Many figures in the national liberation movement have been discussing this increasingly more often. In Syria, for instance, the political activation of the army was caused by the instability of the regime, which proved unable to fulfill the program promised to the people. In these circumstances the officer-patriots used the army as the most organized and dynamic force and rose in opposition to the existing regime. The significance of such events lies in the fact that they are revolutionary, since they lead a country onto the path of social progress in the interests of the broad working masses. But even in this apparently extremely favorable situation the Syrian leaders did not shut themselves within the framework of the army, though they actively relied on it, but proceeded to enlist the working masses into social and political life; they did not balk at creating and arming people's volunteer detachments for defending the revolution.

In a number of progressively developing Asian and African countries great attention has recently been paid to questions of party building and definition of the most acceptable forms of political organization of the masses, those most responsive to the social and national conditions in a given country. In particular, attempts are being made to rebuild the mass parties, which had been created along national-front lines, into vanguard parties that would unite in their ranks the most advanced forces, the most devoted to the cause of the revolution. True, for the time being this process has been moving slowly, and attempts to accelerate it are still very timid. This is especially so if it is compared with the transformations affecting the social and economic field. In discussing the Burmese Socialist Program Party, General Ne Win recently declared, "The party, which must become the political leader of the people, is still not ready to fulfill its leadership role despite its four years of existence."

The process of political organization of the masses unfolds differently in different countries. While some liberated countries adhere to the principle of the one-party system of government, others have coalitions of progressive political parties united in a single national-democratic front. It would be wrong to generalize the experience of one or several countries and automatically transfer it to other countries. But regardless of the form the political organization of the masses takes in a given country, the main point is that the construction of a new society is impossible unless the popular masses take an active part in it.

The army, as the experience of the U.A.R., Burma, Algeria, Syria, and certain other countries showed, has an important role in the national liberation movement and is one of its fighting detachments. But it will be able to perform its mission only given the conditions that it serves the interests of the people, correctly determines its own place in society, and soberly appraises its opportunities for guiding the complex processes of the economic, social, and political development of the liberated states.

6 *Political Development and Political Decay*

SAMUEL P. HUNTINGTON

"Among the laws that rule human societies," de Tocqueville said, "there is one which seems to be more precise and clear than all others. If men are to remain civilized or to become so, the art of associating together must grow and improve in the same ratio in which the equality of conditions is increased."[1] In much of the world today, equality of political participation is growing much more rapidly than is the "art of associating together." The rates of mobilization and participation are high; the rates of organization and institutionalization are low. De Tocqueville's precondition for civilized society is in danger, if it is not already undermined. In these societies, the conflict between mobilization and institutionalization is the crux of

Reprinted by permission of the publisher from *World Politics* XVII (April 1965), pp. 386–430.

I am grateful to the Center for International Affairs, Harvard University, for the support which made this article possible and to Edward C. Banfield, Mather Eliot, Milton J. Esman, H. Field Haviland, Jr., and John D. Montgomery, for their helpful written comments on an earlier draft.

politics. Yet in the fast-growing literature on the politics of the developing areas, political institutionalization usually receives scant treatment. Writers on political development emphasize the processes of modernization and the closely related phenomena of social mobilization and increasing political participation. A balanced view of the politics of contemporary Asia, Africa, and Latin America requires more attention to the "art of associating together" and the growth of political institutions. For this purpose, it is useful to distinguish political development from modernization and to identify political development with the institutionalization of political organizations and procedures. Rapid increases in mobilization and participation, the principal political aspects of modernization, undermine political institutions. Rapid modernization, in brief, produces not political development, but political decay.

Political Development as Modernization

Definitions of political development are legion. Most, however, share two closely related characteristics. First, political development is identified as one aspect of, or as intimately connected with, the broader processes of modernization in society as a whole. Modernization affects all segments of society; its political aspects constitute political development. Indeed, many authors seem to prefer the phrase "political modernization" as more descriptive of their primary concern. Second, if political development is linked with modernization, it is necessarily a broad and complex process. Hence most authors argue that political development must be measured by many criteria. The "multifunction character of politics," Lucian Pye has said, ". . . means that no single scale can be used for measuring the degree of political development."[2] It thus differs from economic development, on the character of which there seems to be more general agreement and which is measurable through fairly precise indices such as per capita national income. Definitions of political development hence tend to itemize a number of criteria. Ward and Rustow list eight characteristics of the modern polity; Emerson has five. Pye identifies four major aspects of political development plus half a dozen additional "factors." Eisenstadt finds four characteristics of political modernization.[3]

The definitions are many and multiple, but, with a few exceptions, the characteristics which they identify with political development are all aspects of the processes of modernization. Four sets of categories recur continuously in the definitions. One set, focusing on the Parsonian pattern variables, can perhaps best be summed up as *rationalization*. This involves movement from particularism to universalism, from diffuseness to specificity, from ascription to achievement, and from affectivity to affective neutrality. In terms of political development, functional differentiation and achievement criteria are particularly emphasized.[4] A second set of characteristics identified with development involves nationalism and *national integration*. Almost all writers recognize the problem of the "crisis of national identity" and the necessity of establishing a firmly delimited ethnic basis for the political community.[5] A developed polity, it is usually assumed, must, with rare exception, be a nation-state. "Nation-building" is a key aspect of political development. A third approach focuses on *democratization:* Pluralism, competitiveness, equalization of power, and similar qualities. "Competitiveness," says Coleman, "is an essential aspect of political modernity. . . ." Hence, "the Anglo-American polities most closely approximate the model of a modern political system. . . ."[6] Frey argues that "the most common notion of political development in intellectual American circles is that of movement towards democracy." He finds this a congenial notion and offers his own definition of political development as "changes in the direction of greater distribution and reciprocity of power. . . ."[7]

Rationalization, integration, and democratization thus commonly appear in definitions of political development. The characteristic of political development or political modernization which is most frequently emphasized, however, is *mobilization,* or *participation.* Modernization, Karl Deutsch has emphasized, involves social mobilization, and "this complex of processes of social change is significantly correlated with major changes in politics." Increases in literacy, urbanization, exposure to mass media, industrialization, and per capita income expand "the politically relevant strata of the population," multiply the demands for government services, and thus stimulate an increase in grovernmental capabilities, a broadening of the elite, increased political participation, and shifts in attention from the local level to the national level.[8] Modernization means mass mobilization, mass mobilization means increased political participation, and increased participation is the key element of political development. Participation distinguishes modern politics from traditional politics. "Traditional society," says Lerner, "is nonparticipant—it deploys people by kinship into communities isolated from each other and from a center. . . ." Modern society, in contrast, is "participant society."[9] The "new world political culture," say Almond and Verba, "will be a political culture of participation. If there is a political revolution going on throughout the world, it is what might be called the participation explosion. In all the new nations of the world the belief that the ordinary man is politically relevant—that he ought to be an involved participant in the political system—is widespread. Large groups of people who have been outside politics are demanding entrance into the political system."[10] Political development, Rustow argues, may be defined as "(1) an increasing national political unity plus (2) a broadening base of political participation. . . ." Similarly, Riggs declares that political development "refers to the process of politicization: increasing participation or involvement of the citizen in state activities, in power calculations, and consequences."[11]

All definitions are arbitrary. These definitions of political development as some combination or permutation of participation, rationalization, democratization, and nation-building are just as legitimate as any other definition. While all definitions may be equally arbitrary and equally legitimate, they do vary greatly, however, in their relevance to particular problems and their usefulness for particular ends. Presumably one major purpose of concepts of political development is to facilitate understanding of the political processes in contemporary Asian, African, and Latin American societies. To be analytically useful, a concept must be precise and relevant. It must also have sufficient generality of application to permit comparative analysis of differing situations. Many approaches to political development suffer from one or more of the following difficulties.

First, the identification of political development with modernization or with factors usually associated with modernization drastically limits the applicability of the concept in both time and space. It is defined in parochial and immediate terms, its relevance limited to modern nation-states or the emergency of modern nation-states. It becomes impossible to speak of a politically developed tribal authority, city-state, feudal monarchy, or bureaucratic empire. Development is identified with one type of political system rather than as a quality which might characterize any type of political system. All systems which are not modern are underdeveloped, including presumably fifth-century Athens, the third-century B.C. Roman republic, the second-century A.D. Roman empire, the Han and Tàng empires in China, or even eighteenth-century America. None of these political systems was modern. Is it also useful to consider them underdeveloped? Would it not be more appropriate to consider development or underdevelopment as a characteristic which might be found in any type of political system? City-states could be developed or underdeveloped; so also could be bureaucratic empires or modern nation-states. This approach would cast additional light on contemporary modernizing societies by furnishing a second set of categories (in addition to the traditional-modern set) for comparing the processes of change in those societies with the processes of change in other types of societies. Such an approach, of course, would also liberate the concept of development from the even more limited identification of it with the Western, constitutional, democratic nation-state.

The second problem with many definitions of political development is the obverse but also the corollary of the first. On the one hand, development is limited to the characteristics of the modern nation-state. On the other, it is also broadened to include almost all politically relevant aspects of the modernization process. It acquires comprehensiveness at the cost of precision. There is a natural tendency to assume that political development is all of a piece, that one "good thing" is compatible with another. In addition, studies of modernization have shown a very high degree of correlation among such indices as literacy, urbanization, media participation, and political participation.[12] Hence, it is easy to assume that a similar correlation exists among the various elements identified as contributing to political development. In fact, however, the four, eight, or twelve criteria of development may or may not have any systematic relation to each other. They may indeed be negatively correlated. There is no particular reason, for instance, why more participation and more structural differentiation should go together; in fact, there is some *a priori* reason to assume that more of one might mean less of the other. If this be the case, two contradictory tendencies (A, −B; −A, B) could both be labeled "political development." The broader the definition of development, moreover, the more inevitable development becomes. The all-encompassing definitions make development seem easy by making it seem inescapable. Development becomes an omnipresent first cause, which explains everything but distinguishes nothing. Almost anything that happens in the "developing" countries—coups, ethnic struggles, revolutionary wars—becomes part of the process of development, however contradictory or retrogressive this may appear on the surface. Political development thus loses its analytical content and acquires simply a geographic one. At the extreme, it becomes synonymous with the political history of Asia, Africa, and Latin America.[13]

Third, many definitions of political development fail to distinguish clearly the empirical relevance of the components going into the definition. Concepts of "developed" and "undeveloped" as ideal types or states of being are confused with concepts of "development" as a process which are, in turn, identified with the politics of the areas commonly called "developing." The line between actuality and aspiration is fogged. Things which are in fact occurring in the "developing" areas become hopelessly intertwined with things which the theorist thinks should occur there. Here again the tendency has been to assume that what is true for the broader processes of social modernization is also true for political changes. Modernization, in some degree, is a fact in Asia, Africa, Latin America: urbanization is rapid; literacy is slowly increasing; industrialization is being pushed; per capita gross national product is inching upward; mass media circulation is expanding; political participation is broadening. All these are facts. In contrast, progress toward many of the other goals identified with political development—democracy, stability, structural differentiation, achievement patterns, national integration—often is dubious at best. Yet the tendency is to think that because modernization is taking place, political development also must be taking place. As a result, many of the sympathetic Western writings about the underdeveloped areas today have the same air of hopeful unreality which characterized much of the sympathetic Western writing about the Soviet Union in the 1920s and 1930s. They are suffused with what can only be described as "Webbism": that is, the tendency to ascribe to a political system qualities which are assumed to be its ultimate goals rather than qualities which actually characterize its processes and functions.[14]

In actuality, only some of the tendencies frequently encompassed in the concept "political development" appear to be characteristic of the "developing" areas. Instead of a trend toward competitiveness and democracy, there has been an "erosion of democracy" and a tendency to autocratic military regimes and one-party regimes. Instead of stability, there have been repeated coups and revolts. Instead of a unifying nationalism and nation-building, there have been repeated ethnic conflicts and civil wars. Instead of institutional rationalization and differentiation, there has frequently been a decay of the administrative organizations inherited from the colonial era and a weakening and disruption of the political organizations developed during the struggle for independence.[15] Only the concept of political development as mobilization and participation appears to be generally applicable to the "developing" world. Rationalization, competitiveness, and nation-building, in contrast, seem to have only a dim relation to reality.

This gap between theory and reality suggests a fourth difficulty in many concepts of political development. They are usually one-way concepts. Little or no provision is made for their reversibility. If political development is thought to involve the mobilization of people into politics, account should also be taken of the possibility that political de-development can take place and people can be demobilized out of politics. Structural differentiation may occur, but so also may structural homogenization. National disintegration is a phenomenon as much as national integration. A concept of political development should be reversible. It should define both political development and the circumstances under which political decay is encouraged.

The failure to think of political development as a reversible process apparently stems from two sources. Insofar as development is identified with modernization, many aspects of modernization do appear to be practically irreversible. Urbanization is not likely to give way to ruralization. Increases in literacy are not normally followed by sharp declines. Capital once invested in factories or power plants stays invested. Even increases in per capita gross national product are, more often than not, permanent, except for minor dips or destruction caused by war or natural catastrophe. With varying slopes, with hesitancy in some sectors but with strength and steady progress in others, virtually all the indices of modernization progress steadily upward on the charts. But political changes have no such irreversibility.

In other instances, one feels that an underlying commitment to the theory of progress is so overwhelming as to exclude political decay as a possible concept. Political decay, like thermonuclear war, becomes unthinkable. Almond, for instance, measures not just political development but *political change* by "the acquisition by a political system of some new capability."[16] The specific capabilities he has in mind are those for national intergration, international accommodation, political participation, and welfare distribution. Before the Renaissance, Almond argues, political systems "acquired and lost capabilities . . . in anything but a unilinear, evolutionary way." Modernization, however, reduces "the independence of man's political experiments." Change is "far from unilinear," but it is toward "the emergence of world culture." Surely, however, modern and modernizing states can change by losing capabilities as well as by gaining them. In addition, a gain in any one capability usually involves costs in others. A theory of political development needs to be mated to a theory of political decay. Indeed, as was suggested above, theories of instability, corruption, authoritarianism, domestic violence, institutional decline, and political disintegration may tell us a lot more about the "developing" areas than their more hopefully defined opposites.

Political Development as Institutionalization

There is thus much to be gained (as well as something to be lost) by conceiving of political development as a process independent of, although obviously affected by, the process of modernization. In view of the crucial importance of the relationship between mobilization and participation, on the one hand, and the growth of political organizations, on the other, it is useful for many purposes to define political development as the institutionalization of political organizations and procedures. This concept liberates development from modernization. It can be applied to the analysis of political systems of any sort, not just modern ones. It can be defined in reasonably precise ways which are at least theoretically capable of measurement. As a concept, it does not suggest that movement is likely to be in only one direction: institutions, we know, decay and dissolve as well as grow and mature. Most significantly, it focuses attention on the reciprocal interaction between the on-going social processes of modernization, on the one hand, and the strength, stability, or weakness of political structures, traditional, transitional, or modern, on the other.[17]

The strength of political organizations and procedures varies with their *scope of support* and their *level of institutionalization.* Scope refers simply to the extent to which the political organizations and procedures encompass activity in the society. If only a small upper-class group belongs to political organizations and behaves in terms of a set of procedures, the scope is limited. If, on the other hand, a large segment of the population is politically organized and follows the political procedures, the scope is broad. Institutions are stable, valued, recurring patterns of behavior. Organizations and procedures vary in their degree of institutionalization. Harvard University and the newly opened suburban high school are both organizations, but Harvard is much more of an institution than is the high school. The seniority system in Congress and President Johnson's select press conferences are both procedures, but seniority is much more institutionalized than are Mr. Johnson's methods of dealing with the press. Institutionalization is the process by which organizations and procedures acquire value and stability. The level of institutionalization of any political system can be defined by the adaptability, complexity, autonomy, and coherence of its organizations and procedures. So also, the level of institutionalization of any particular organization or procedure can be measured by its adaptability, complexity, autonomy, and coherence. If these criteria can be identified and measured, political systems can be compared in terms of their levels of institutionalization. Furthermore, it will be possible to measure increases and decreases in the institutionalization of particular organizations and procedures within a political system.

Adaptability-Rigidity

The more adaptable an organization or procedure is, the more highly institutionalized it is; the less adaptable and more rigid it is, the lower its level of institutionalization. Adaptability is an acquired organizational characteristic. It is, in a rough sense, a function of environmental challenge and age. The more challenges which have arisen in its environment and the greater its age, the more adaptable it is. Rigidity is more characteristic of young organizations than of old ones. Old organizations and procedures, however, are not necessarily adaptable if they have existed in a static environment. In addition, if over a period of time an organization has developed a set of responses for dealing effectively with one type of problem, and if it is then confronted with an entirely different type of problem requiring a different response, the organization may well be a victim of its past successes and be unable to adjust to the new challenge. In general, however, the first hurdle is the biggest one. Success in adapting to one environmental challenge paves the way for successful adaptation to subsequent environmental challenges. If, for instance, the probability of successful adjustment to the first challenge is 50 percent, the probability of successful adjustment to the second challenge might be 75 percent, to the third challenge 87½ percent, to the fourth 93¾ percent, and so on. Some changes in environment, moreover, such as changes in personnel, are inevitable for all organizations. Other changes in environment may be produced by the organization itself; if, for instance, it successfully completes the task which it was originally created to accomplish. So long as it is recognized that environments can differ in the challenges which they pose to organizations, the adaptability of an organization can in a rough sense be measured by its age. Its age, in turn, can be measured in three ways.

One is simply chronological: the longer an organization or procedure has been in existence, the higher the level of institutionalization. The older an organization is, the more likely it is to continue to exist through any specified future time period. The probability that an organization which is one hundred years old will survive one additional year, it might be hypothesized, is perhaps one hundred times greater than the probability that an organization one year old will survive one additional year. Political institutions are thus not created overnight. Political development, in this sense, is slow, particularly when compared with the seemingly much more rapid pace of economic development. In some instances, particular types of experience may substitute for time: fierce conflict or other serious challenges may transform organizations into institutions much more rapidly than normal circumstances. But such intensive experiences are rare, and even with such experiences time is still required. "A major party," Ashoka Mehta has observed, in commenting on why communism is helpless in India, "cannot be created in a day. In China a great party was forged by the revolution. Other major parties can be or are born of revolutions in other countries. But it is simply impossible, through normal channels, to forge a great party, to reach and galvanize millions of men in half a million villages."[18]

A second measure of adaptability is generational age. So long as an organization still has its first set of leaders, so long as a procedure is still performed by those who first performed it, its adaptability is still in doubt. The more often the organization has surmounted the problem of peaceful succession and replaced one set of leaders with another, the more highly institutionalized it is. In considerable measure, of course, generational age is a function of chronological age. But political parties and governments may continue for decades under the leadership of one generation. The founders of organizations—whether parties, governments, or business corporations—are often young. Hence the gap between chronological age and generational age is apt to be greater in the early history of an organization than later in its career. This gap produces tensions between the first leaders of the organization and the next generation immediately behind them, which can look forward to a lifetime in the shadow of the first generation. In the middle of the 1960s the Chinese Communist Party was forty-five years old, but in large part it was still led by its first generation of leaders. An organization may also change leadership without changing generations of leadership. One generation differs from another in terms of its formative experiences. Simple replacement of one set of leaders by another, that is, surmounting a succession crisis, counts for something in terms of institutional adaptability, but it is not as significant as a shift in leadership generations, that is, the replacement of one set of leaders by another set with significantly different organizational experiences. The shift from Lenin to Stalin was an intrageneration succession; the shift from Stalin to Khrushchev was an inter-generation succession.

Third, organizational adaptability can be measured in functional terms. An organization's functions, of course, can be defined in an almost infinite number of ways. (This is a major appeal and a major limitation of the functional approach to organizations.) Usually an organization is created to perform one particular function. When that function is no longer needed, the organization faces a major crisis. It either finds a new function or reconciles itself to a lingering death. An organization which has adapted itself to changes in its environment and has survived one or more changes in its principal functions is more highly institutionalized than one which has not. Not functional specificity but functional adaptability is the true measure of a highly developed organization. Institutionalization makes the organization more than simply an instrument to achieve certain purposes.[19] Instead its leaders and members come to value it for its own sake, and it develops a life of its own quite apart from the specific functions it may perform at any given time. The organization triumphs over its function.

Organizations and individuals thus differ significantly in their cumulative capacity to adapt to changes. Individuals usually grow up through childhood and adolescence without deep commitments to highly specific functions. The process of commitment begins in late adolescence. As the individual becomes more and more committed to the performance of certain functions, he finds it increasingly difficult to change those functions and to unlearn the responses which he has acquired to meet environmental changes. His personality has been formed; he has become "set in his ways." Organizations, on the other hand, are usually created to perform very specific functions. When the organization confronts a changing environment, it must, if it is to survive, weaken its commitment to its original functions. As the organization matures, it becomes "unset" in its ways.

In practice, organizations vary greatly in their functional adaptability. The YMCA, for instance, was founded in the mid-nineteenth century as an evangelical organization to convert the single young men who, during the early years of industrialization, were migrating in great numbers to the cities. With the decline in need for this function, the Y successfully adjusted to the performance of many other "general service" functions broadly related to the legitimizing goal of "character development." Concurrently, it broadened its membership base to include first nonevangelical Protestants, then Catholics, then Jews, then old men as well as young, and then women as well as men![20] As a result, the organization has prospered although its original functions disappeared with the dark satanic mills. Other organizations, such as the WCTU and the Townsend Movement, have had greater difficulty in adjusting to a changing environment. The WCTU "is an organization in retreat. Contrary to the expectations of theories of institutionalization, the movement has not acted to preserve organizational values at the expense of past doctrine."[21] The Townsend Movement has been torn between those who wish to remain loyal to the original function and those who put organizational imperatives first. If the latter are successful, "the dominating orientation of leaders and members shifts *from the implementation of the values the organization is taken to represent* (by leaders, members, and public alike), *to maintaining the organizational structure as such,* even at the loss of the organization's central mission."[22] The conquest of polio posed a similar acute crisis for the National Foundation for Infantile Paralysis. The original goals of the organization were highly specific. Should the organization dissolve when these goals were achieved? The dominant opinion of the volunteers was that the organization should continue. "We can fight polio," said one town chairman, "if we can organize people. If we can organize people like this we can fight anything." Another felt that: "Wouldn't it be a wonderful story to get polio licked, and then go on to something else and get that licked and then go on to something else? It would be a challenge, a career."[23]

The problems of functional adaptability are not much different for political organizations. A political party gains in functional age when it shifts its function from the representation of one constituency to the representation of another; it also gains in functional age when it shifts from opposition to government. A party which is unable to change constituencies or to acquire power is less of an institution than one which is able to make these changes. A nationalist party whose function has been the promotion of independence from colonial rule faces a major crisis when it achieves its goal and has to adapt itself to the somewhat different function of governing a country. It may find this functional transition so difficult that it will, even after independence, continue to devote a large portion of its efforts to fighting colonialism. A party which acts this way is less of an institution than one, like the Congress Party, which after achieving independence drops its anticolonialism and quite rapidly adapts itself to the tasks of governing. Industrialization has been a major function of the Communist Party of the Soviet Union. A major test of the institutionalization of the Communist Party will be its success in developing new functions now that the major industrializing effort is behind it. A governmental organ which can successfully adapt itself to changed functions, such as the British Crown in the eighteenth and nineteenth centuries, is more of an institution than one which cannot, such as the French monarchy in the same period.

Complexity-Simplicity

The more complicated an organization is, the more highly institutionalized it is. Complexity may involve both multiplication of organizational subunits, hierarchically and functionally, and differentiation of separate types of organizational subunits. The greater the number and variety of subunits, the greater the ability of the organization to secure and maintain the loyalties of its members. In addition, an organization which has many purposes is better able to adjust itself to the loss of any one purpose than an organization which has only one purpose. The diversified corporation is obviously less vulnerable than that which produces one product for one market. The differentiation of subunits within an organization may or may not be along functional lines. If it is functional in character, the subunits themselves are less highly institutionalized than the whole of which they are a part. Changes in the functions of the whole, however, are fairly easily reflected by changes in the power and roles of its subunits. If the subunits are multifunctional, they have greater institutional strength, but they may also, for that very reason, contribute less flexibility to the organization as a whole. Hence, a political system with parties of "social integration," in Neumann's terms, has less institutional flexibility than one with parties of "individual representation."[24]

Relatively primitive and simple traditional political systems are usually overwhelmed and destroyed in the modernization process. More complex traditional systems are more likely to adapt to these new demands. Japan, for instance, was able to adjust its traditional political institutions to the modern world because of their relative complexity. For two and a half centuries before 1868, the emperior had reigned and the Tokugawa shogun had ruled. The stability of the political order, however, did not depend solely on the stability of the shogunate. When the authority of the shogunate decayed, another traditional institution, the emperor, was available to become the instrument of the modernizing samurai. The collapse of the shogun involved not the overthrow of the political order but the "restoration" of the emperor.

The simplest political system is that which depends on one individual. It is also, of course, the least stable. Tyrannies, Aristotle pointed out, are virtually all "quite short-lived."[25] A political system with several different political institutions, on the other hand, is much more likely to adapt. The needs of one age may be met by one set of institutions, the needs of the next by a different set. The system possesses within itself the means of its own renewal and adaptation. In the American system, for instance, president, Senate, House of Representatives, Supreme Court, and state governments have played different roles at different times in history. As new problems arise, the initiative in dealing with them may be taken first by one institution, then by another. In contrast, the French system of the Third and Fourth Republics centered authority in the National Assembly and the national bureaucracy. If, as was frequently the case, the Assembly was too divided to act and the bureaucracy lacked the authority to act, the system was unable to adapt to environmental changes and to deal with new policy problems. When in the 1950s the Assembly was unable to handle the dissolution of the French Empire, there was no other institution, such as an independent executive, to step into the breach. As a result, an extraconstitutional force, the military, intervened in politics, and in due course a new institution, the de Gaulle presidency, was created which was able to handle the problem. "A state without the means of some change," Burke observed of an earlier French crisis, "is without the means of its conservation."[26]

The classical political theorists, preoccupied as they were with the problem of stability, arrived at similar conclusions. The simple forms of government were most likely to degenerate; the "mixed state" was more likely to be stable. Both Plato and Aristotle suggested that the most practical state was the "polity" combining the institutions of democracy and oligarchy. A "constitutional system based absolutely, and at all points," Aristotle argued, "on either the oligarchical or the democratic conception of equality is a poor sort of thing. The facts are evidence enough: constitutions of this sort never endure." A "constitution is better when it is composed of more numerous elements."[27] Such a constitution is more likely to head off sedition and revolution. Polybius and Cicero elaborated this idea more explicitly. Each of the "good" simple forms of government—kingship, aristocracy, and democracy—is likely to degenerate into its perverted counterpart—tyranny, oligarchy, and mobocracy. Instability and degeneration can be avoided only by combining elements from all the good forms into a mixed state. Complexity produces stability. "The simple governments," Burke echoed two thousand years later, "are fundamentally defective, to say no worse of them."[28]

Autonomy-Subordination

A third measure of institutionalization is the extent to which political organizations and procedures exist independently of other social groupings and methods of behavior. How well is the political sphere differentiated from other spheres? In a highly developed political system, political organizations have an integrity which they lack in less developed systems. In some measure, they are insulated from the impact of nonpolitical groups and procedures. In less developed political systems, they are highly vulnerable to outside influences.

At its most concrete level, autonomy involves the relations between social forces, on the one hand, and political organizations, on the other. Social forces include the groupings of men for social and economic activities: families, clans, work groups, churches, ethnic and linguistic groupings. Political institutionalization, in the sense of autonomy, means the development of political organizations and procedures which are not simply expressions of the interests of particular social groups. A political organization which is the instrument of a social group—family, clan, class—lacks autonomy and institutionalization. If the state, in the traditional Marxist claim, is really the "executive committee of the bourgeoisie," then it is not much of an institution. A judiciary is independent to the extent that it adheres to distinctly judicial norms and to the extent that its perspectives and behavior are independent of those of other political institutions and social groupings. As with the judiciary, the autonomy of political institutions is measured by the extent to which they have their own interests and values distinguishable from those of other social forces. As with the judiciary, the autonomy of political institutions is likely to be the result of competition among social forces. A political party, for instance, which expresses the interests of only one group in society—whether labor, business, or farmers—is less autonomous than one which articulates and aggregates the interests of several social groups. The latter type of party has a clearly defined existence apart from particular social forces. So also with legislatures, executives, and bureaucracies. Political procedures, like political organizations, also have varying degrees of autonomy. A highly developed political system has procedures to minimize, if not to eliminate, the role of violence in the system and to restrict to explicitly defined channels the influence of wealth in the system. To the extent that political officials can be toppled by a few soldiers or influenced by a few dollars, the organizations and procedures lack autonomy. Political organizations and procedures which lack autonomy are, in common parlance, said to be corrupt.

Political organizations and procedures which are vulnerable to nonpolitical influences from within the society are also usually vulnerable to influences from outside the society. They are easily penetrated by agents, groups, and ideas from other political systems. Thus, a coup d'état in one political system may easily "trigger" a coup d'état by similar groups in other less developed political systems.[29] In some instances, apparently, a regime can be overthrown by smuggling into the country a few agents and a handful of weapons. In other instances, a regime may be overthrown by the exchange of a few words and a few thousand dollars between a foreign ambassador and some disaffected colonels. The Soviet and American governments presumably spend substantial sums attempting to bribe high officials of less well-insulated political systems which they would not think of wasting in attempting to influence high officials in each other's political system.

In every society affected by social change, new groups arise to participate in politics. Where the political system lacks autonomy, these groups gain entry into politics without becoming identified with the established political organizations or acquiescing in the established political procedures. The political organizations and procedures are unable to stand up against the impact of a new social force. Conversely, in a developed political system, the autonomy of the system is protected by mechanisms which restrict and moderate the impact of new groups. These mechanisms either slow down the entry of new groups into politics or, through a process of political socialization, impel changes in the attitudes and behavior of the most politically active members of the new group. In a highly institutionalized political system, the most important positions of leadership can normally be achieved only by those who have served an apprenticeship in less important positions. The complexity of a political system contributes to its autonomy by providing a variety of organizations and positions in which individuals are prepared for the highest offices. In a sense, the top positions of leadership are the inner core of the political system; the less powerful positions, the peripheral organizations, and the semipolitical organizations are the filters through which individuals desiring access to the core must pass. Thus the political system assimilates new social forces and new personnel without sacrificing its institutional integrity. In a political system which lacks such defenses, new men, new viewpoints, new social groups may replace each other at the core of the system with bewildering rapidity.

Coherence-Disunity

The more unified and coherent an organization is, the more highly institutionalized it is; the greater the disunity of the organization, the less its institutionalization. Some measure of consensus, of course, is a prerequisite for any social group. An effective organization requires, at a minimum, substantial consensus on the functional boundaries of the group and on the procedures for resolving disputes on issues which come up within those boundaries. The consensus must extend to those active in the system. Nonparticipants or those only sporadically and marginally participant in the system do not have to share the consensus and usually, in fact, do not share it to the same extent as the participants.[30] In theory, an organization can be autonomous without being coherent and coherent without being autonomous. In actuality, however, the two are often closely linked together. Autonomy becomes a means to coherence, enabling the organization to develop an esprit and style which become distinctive marks of its behavior. Autonomy also prevents the intrusion of disruptive external forces, although, of course, it does not protect against disruption from internal sources. Rapid or substantial expansions in the membership of an organization or in the participants in a system tend to weaken coherence. The Ottoman Ruling Institution, for instance, retained its vitality and coherence as long as admission was restricted and recruits were "put through an elaborate education, with selection and specialization at every stage." The Institution perished when "everybody pressed in to share its privileges. . . . Numbers were increased; discipline and efficiency declined."[31]

Unity, esprit, morale, and discipline are needed in governments as well as in regiments. Numbers, weapons, and strategy all count in war, but major deficiencies in any one of those may still be counterbalanced by superior coherence and discipline. So also in politics. The problems of creating coherent political organizations are more difficult but not fundamentally different from those involved in the creation of coherent military organizations. "The sustaining sentiment of a military force," David Rapoport has argued, "has much in common with that which cements any group of men engaged in politics—the willingness of most individuals to bridle private or personal impulses for the sake of general social objectives. Comrades must trust each other's ability to resist the innumerable temptations that threaten the group's solidarity; otherwise, in trying social situations the desire to fend for oneself becomes overwhelming."[32] The capacities for coordination and discipline are crucial to both war and politics, and historically societies which have been skilled at organizing the one have also been adept at organizing the other. "The relationship of efficient social organization in the arts of peace and in the arts of group conflict," one anthropologist has observed, "is almost absolute, whether one is speaking of civilization or subcivilization. Successful war depends upon team work and consensus, both of which require command and discipline. Command and discipline, furthermore, can eventually be no more than symbols of something deeper and more real than they themselves."[33] Societies, such as Sparta, Rome, and Britain, which have been admired by their contemporaries for the authority and justice of their laws have also been admired for the coherence and discipline of their armies. Discipline and development go hand in hand.

One major advantage of studying development in terms of mobilization and participation is that they are measurable. Statistics are readily available for urbanization, literacy, mass media exposure, and voting. Hence, comparisons are easily made between countries and between different stages of the same country. What about institutionalization? Are the criteria of adaptability, complexity, autonomy, and coherence also measurable? Quite obviously the difficulties are greater. The UN has not conveniently collected in its *Statistical Yearbook* data on the political institutionalization of its members. Nonetheless, no reason exists why with a little imagination and effort sufficient information could not be collected to make meaningful comparisons of the levels of political institutionalization of different countries or of the same country at different times. Adaptability can be measured by chronological age, leadership successions, generational changes, and functional changes. Complexity can be measured by the number and diversity of organizational subunits and by the number and diversity of functions performed by the organizations. Autonomy is perhaps the most difficult of the criteria to pin down: it can, however, be measured by the distinctiveness of the norms and values of the organization compared with those of other groups, by the personnel controls (in terms of cooptation, penetration, and purging) existing between the organization and other groups, and by the degree to which the organization controls its own material resources. Coherence may be measured by the ratio of contested successions to total successions, by the cumulation or noncumulation of cleavages among leaders and members, by the incidence of overt alienation and dissent within the organization, and, conceivably, by opinion surveys of the loyalties and preferences of organization members.

Experience tells us that levels of institutionalization differ. Measuring that difference may be difficult, but it is not impossible. Only by measuring institutionalization will we be able to buttress or disprove hypotheses about the relation between social, economic, and demographic changes, on the one hand, and variations in political structure, on the other.

Mobilization versus Institutionalization: Public Interests, Degeneration, and the Corrupt Polity

Mobilization and Institutionalization

Social mobilization and political participation are rapidly increasing in Asia, Africa, and Latin America. These processes, in turn, are directly responsible for the deterioration of political institutions in these areas. As Kornhauser has conclusively demonstrated for the Western world, rapid industrialization and urbanization create discontinuities which give rise to mass society. "The *rapid* influx of large numbers of people into *newly* developing urban areas invites mass movements."[34] In areas and industries with very rapid industrial growth, the creation and institutionalization of unions lag, and mass movements are likely among the workers. As unions are organized, they are highly vulnerable to outside influences in their early stages. "The rapid influx of large numbers of people into a new organization (as well as a new area) provides opportunities for mass-oriented elites to penetrate the organization. This is particularly true during the formative periods of organizations, for at such times external constraints must carry the burden of social control until the new participants have come to internalize the values of the organization."[35]

So also in politics. Rapid economic growth breeds political instability.[36] Political mobilization, moreover, does not necessarily require the building of factories or even movement to the cities. It may result simply from increases in communications, which can stimulate major increases in aspirations that may be only partially, if at all, satisfied. The result is a "revolution of rising frustrations."[37] Increases in literacy and education may bring more political instability. By Asian standards, Burma, Ceylon, and the Republic of Korea are all highly literate, but no one of them is a model of political stability. Nor does literacy necessarily stimulate democracy: with roughly 75 percent literacy, Cuba was the fifth most literate country in Latin America (ranking behind Argentina, Uruguay, Chile, and Costa Rica), but the first to go Communist; so also Kerala, with one of the highest literacy rates in India, was the first Indian state to elect a Communist government.[38] Literacy, as Daniel Lerner has suggested, "may be dysfunctional—indeed a serious impediment—to modernization in the societies now seeking (all too rapidly) to transform their institutions."[39]

Increased communication may thus generate demands for more "modernity" than can be delivered. It may also stimulate a reaction against modernity and activate traditional forces. Since the political arena is normally dominated by the more modern groups, it can bring into the arena new, antimodern groups and break whatever consensus exists among the leading political participants. It may also mobilize minority ethnic groups who had been indifferent to politics but who now acquire a self-consciousness and divide the political system along ethnic lines. Nationalism, it has often been assumed, makes for national integration. But in actuality, nationalism and other forms of ethnic consciousness often stimulate political disintegration, tearing apart the body politic.

Sharp increases in voting and other forms of political participation can also have deleterious effects on political institutions.

In Latin America since the 1930s, increases in voting and increases in political instability have gone hand in hand. "Age requirements were lowered, property and literacy requirements were reduced or discarded, and the unscrubbed, unschooled millions on the farms were enfranchised in the name of democracy. They were swept into the political life of the republics so rapidly that existing parties could not absorb many of them, and they learned little about working within the existing political system."[40] The personal identity crises of the elites, caught between traditional and modern cultures, may create additional problems: "In transitional countries the political process often has to bear to an inordinate degree the stresses and strains of people responding to personal needs and seeking solutions to intensely personal problems."[41] Rapid social and economic change calls into question existing values and behavior patterns. It thus often breeds personal corruption. In some circumstances this corruption may play a positive role in the modernizing process, enabling dynamic new groups to get things done which would have been blocked by the existing value system and social structure. At the same time, however, corruption undermines the autonomy and coherence of political institutions. It is hardly accidental that in the 1870s and 1880s a high rate of American economic development coincided with a low point in American governmental integrity.[42]

Institutional decay has become a common phenomenon of the modernizing countries. Coups d'etat and military interventions in politics are one index of low levels of political institutionalization: they occur where political institutions lack autonomy and coherence. According to one calculation, eleven of twelve modernizing states outside Latin America which were independent before World War II experienced coups d'etat or attempted coups after World War II. Of twenty states which became independent between World War II and 1959, fourteen had coups or coup attempts by 1963. Of twenty-four states which became independent between 1960 and 1963, seven experienced coups or attempted coups before the end of 1963.[43] Instability in Latin

America was less frequent early in the twentieth century than it was in the middle of the century. In the decade from 1917 to 1927, military men occupied the presidencies of the twenty Latin American republics 28.7 percent of the time; in the decade from 1947 to 1957, military men were presidents 45.5 percent of the time.[44] In the 1930s and 1940s in countries like Argentina and Colombia, military intervention in politics occurred for the first time in this century. Seventeen of the twenty Latin American states experienced coups or coup attempts between 1945 and 1964, only Chile, Mexico, and Uruguay having clean records of political stability.

In many states the decline of party organizations is reflected in the rise of charismatic leaders who personalize power and weaken institutions which might limit that power. The increasing despotism of Nkrumah, for instance, was accompanied by a marked decline in the institutional strength of the Convention People's Party. In Turkey, Pakistan, and Burma, the Republican People's Party, Muslim League, and AFPFL deteriorated and military intervention eventually ensued. In party organizations and bureaucracies, marked increases in corruption often accompanied significant declines in the effectiveness of governmental services. Particularistic groups—tribal, ethnic, religious—frequently reasserted themselves and further undermined the authority and coherence of political institutions. The legitimacy of postcolonial regimes among their own people was often less than that of the colonial regimes of the Europeans. Economists have argued that the gap between the level of economic well-being of the underdeveloped countries and that of highly developed countries is widening as the absolute increases and even percentage increases of the latter exceed those of the former. Something comparable and perhaps even more marked is occurring in the political field. The level of political institutionalization of the advanced countries has, with a few exceptions such as France, remained relatively stable. The level of political institutionalization of most other countries has declined. As a result, the political gap between them has broadened. In terms of institutional strength, many if not most of the new states reached their peak of political development at the moment of independence.

The differences which may exist in mobilization and institutionalization suggest four ideal types of politics (see Table 1). Modern, developed, civic polities (the United States, the Soviet Union) have high levels of both mobilization and institutionalization. Primitive polities (such as Banfield's backward society) have low levels of both. Contained polities are highly institutionalized but have low levels of mobilization and participation. The dominant political institutions of contained polities may be either traditional (e.g., monarchies) or modern (e.g., political parties). If they are the former, such polities may well confront great difficulties in adjusting to rising levels of social mobilization. The traditional institutions may wither or collapse, and the result would be a corrupt polity with a high rate of participation but a low level of institutionalization. In the corrupt society, politics is, in Macaulay's phrase, "all sail and no anchor."[45] This type of polity characterizes much, if not most, of the modernizing world. Many of the more advanced Latin American countries, for instance, have achieved comparatively high indices of literacy, per capita national income, and urbanization. But their politics remains notably underdeveloped. Distrust and hatred have produced a continuing low level of political institutionalization. "There is no good faith in America, either among men or among nations," Bolivar once lamented. "Treaties are paper, constitutions books, elections battles, liberty anarchy, and life a torment. The only thing one can do in America is emigrate."[46] Over a century later, the same complaint was heard: "We are not, or do not represent a respectable nation . . . not because we are poor, but because we are disorganized," argued

TABLE 1: *Types of Political Systems*

Social Mobilization	*Political Institutionalization* *High*	*Low*
High	Civic	Corrupt
Low	Contained	Primitive

an Ecuadorian newspaper. "With a politics of ambush and of permanent mistrust, one for the other, we . . . cannot properly organize a republic . . . and without organization we cannot merit or attain respect from other nations."[47] So long as a country like Argentina retains a politics of coup and countercoup and a feeble state surrounded by massive social forces, it cannot be considered politically developed, no matter how urbane and prosperous and educated are its citizens.

In reverse fashion, a country may be politically highly developed, with modern political institutions, while still very backward in terms of modernization. India, for instance, is typically held to be the epitome of the underdeveloped society. Judged by the usual criteria of modernization, it was at the bottom of the ladder during the 1950s: per capita GNP of $72, 80 percent illiterate, over 80 percent of the population in rural areas, 70 percent of the work force in agriculture, a dozen major languages, deep caste and religious differences. Yet in terms of political institutionalization, India was far from backward. Indeed, it ranked high not only in comparison with other modernizing countries in Asia, Africa, and Latin America but also in comparison with many much more modern European countries. A well-developed political system has strong and distinct institutions to perform both the "input" and the "output" functions of politics. India entered independence with not only two organizations, but two highly developed—adaptable, complex, autonomous, and coherent—institutions ready to assume primary responsibility for these functions. The Congress Party, founded in 1885, was one of the oldest and best-organized political parties in the world; the Indian Civil Service, dating from the early nineteenth century, has been appropriately hailed as "one of the greatest administrative systems of all time."[48] The stable, effective, and democratic government of India during the first fifteen years of independence rested far more on this institutional inheritance than it did on the charisma of Nehru. In addition, the relatively slow pace of modernization and social mobilization in India did not create demands and strains which the Party and the bureaucracy were unable to handle. So long as these two organizations maintain their institutional strength, it is ridiculous to think of India as politically underdeveloped, no matter how low her per capita income or how high her illiteracy rate.

Almost no other country which became independent after World War II was institutionally as well prepared as India for self-government. In countries like Pakistan and the Sudan, institutional evolution was unbalanced; the civil and military bureaucracies were more highly developed than the political parties, and the military had strong incentives to move into the institutional vacuum on the input side of the political system and to attempt to perform interest aggregation functions. This pattern, of course, has also been common in Latin America. In countries like Guatemala, El Salvador, Peru, and Argentina, John J. Johnson has pointed out, the military is "the country's best organized institution and is thus in a better position to give objective expression to the national will" than are parties or interest groups.[49] In a very different category is a country like North Vietnam, which fought its way into independence with a highly disciplined political organization but which was distinctly weak on the administrative side. The Latin American parallel here would be Mexico, where, as Johnson puts it, "not the armed forces but the PRI is the best organized institution, and the party rather than the armed forces has been the unifying force at the national level." In yet a fourth category are those unfortunate states, such as the Congo, which were born with neither political nor administrative institutions. Many of these new states deficient at independence in one or both types of institutions have also been confronted by high rates of social mobilization and rapidly increasing demands on the political system (see Table 2).

TABLE 2: *Institutional Development at Moment of Independence*

Input Institutions	*Output Institutions*	
	High	*Low*
High	India	North Vietnam
Low	Sudan	Congo

Political Institutions and Public Interests

A society with weak political institutions lacks the ability to curb the excesses of personal and parochial desires. Politics is a Hobbesian world of unrelenting competition among social forces—between man and man, family and family, clan and clan, region and region, class and class—a competition unmediated by more comprehensive political organizations. The "amoral familism" of Banfield's village has its counterparts in amoral clanism, amoral groupism, and amoral classism. Without strong political institutions, society lacks the means of defining and realizing its common interests. The capacity to create political institutions is the capacity to create public interests.

Traditionally the public interest has been approached in three ways.[50] It has been identified either with abstract, substantive ideal values and norms such as natural law, justice, or right reason; or with the specific interest of a particular individual ("L'état, c'est moi"), group, class (Marxism), or majority; or with the result of a competitive process among individuals (classic liberalism) or groups (Bentleyism). The problem in all these approaches is to arrive at a definition which is concrete rather than nebulous and general rather than particular. Unfortunately, in most cases what is concrete lacks generality and what is general lacks concreteness. One partial way out of the problem is to define the public interest in terms of the concrete interests of the governing institutions. A society with highly institutionalized governing organizations and procedures is, in this sense, more able to articulate and achieve its public interests. "Organized (institutionalized) political communities," as Friedrich argues, "are *better adapted* to reaching decisions and developing policies than unorganized communities."[51] The public interest, in this sense, is not something which exists a priori in natural law or the will of the people. Nor is it simply whatever results from the political process. Rather it is whatever strengthens governmental institutions. The public interest is the interest of public institutions. It is something which is created and brought into existence by the institutionalization of government organizations. In a complex political system, many governmental organizations and procedures represent many different aspects of the public interest. The public interest of a complex society is a complex matter.

We are accustomed to think of our primary governing institutions as having representative functions—that is, as expressing the interests of some other set of groups (their constituency). Hence, we tend to forget that governmental institutions have interests of their own. These interests not only exist; they are also reasonably concrete. The questions, "What is the interest of the presidency? What is the interest of the Senate? What is the interest of the House of Representatives? What are the interests of the Supreme Court?" are difficult but not completely impossible to answer. The answers would furnish a fairly close approximation of the "public interest" of the United States. Similarly, the public interest of Great Britain might be approximated by the specific institutional interests of the Crown, Cabinet, and Parliament. In the Soviet Union, the answer would involve the specific institutional interests of the Presidium, Secretariat, and Central Committee of the Communist Party.

Institutional interests differ from the interests of individuals who are in the institutions. Keynes's percipient remark that "In the long run, we are all dead" applies to individuals, not institutions. Individual interests are necessarily short-run interests. Institutional interests, however, exist through time: the proponent of the institution has to look to its welfare through an indefinite future. This consideration often means a limiting of immediate goals. The "true policy," Aristotle remarked, "for democracy and oligarchy alike, is not one which ensures the greatest possible amount of either, but one which will ensure the longest possible life for both."[52] The official who attempts to maximize power or other values in the short run often weakens his institution in the long run. Supreme Court justices may, in terms of their immediate individual desires, wish to declare an act of Congress unconstitutional. In deciding whether it is in the public interest to do so, however, presumably one question they should ask themselves is whether it is in the long-term institutional interest of the Supreme Court for them to do so. Judicial statesmen are those who, like John Marshall in *Marbury vs. Madison,* maximize the institutional power of the Court in such a way that it is impossible for either the president or Congress to challenge it. In contrast, the Supreme Court justices of the 1930s came very close to expanding their immediate influence at the expense of the long-term interests of the Court as an institution.

The phrase "What's good for General Motors is good for the country" contains at least a partial truth. "What's good for the presidency is good for the country," however, contains more truth. Ask any reasonably informed group of Americans to identify the five best presidents and the five worst presidents. Then ask them to identify the five strongest presidents and the five weakest presidents. If the identification of strength with goodness and weakness with badness is not 100 percent, it will almost certainly not be less than 80 percent. Those presidents—Jefferson, Lincoln, the Roosevelts, Wilson—who expanded the powers of their office are hailed as the beneficient promoters of the public welfare and national interest. Those presidents, such as Buchanan, Grant, Harding, who failed to defend the power of their institution against other groups are also thought to have done less good for the country. Institutional interest coincides with public interest. The power of the presidency is identified with the good of the polity.

The public interest of the Soviet Union is approximated by the institutional interests of the top organs of the Communist Party: "What's good for the Presidium is good for the Soviet Union." Viewed in these terms, Stalinism can be defined as a situation in which the personal interests of the ruler take precedence over the institutionalized interests of the Party. Beginning in the late 1930s Stalin consistently weakened the Party. No Party Congress was held between 1939 and 1952. During and after World War II the Central Committee seldom met. The Party secretariat and Party hierarchy were weakened by the creation of competing organs. Conceivably this process could have resulted in the displacement of one set of governing institutions by another, and some American experts and some Soviet leaders did think that governmental organizations rather than Party organizations would become the ruling institutions in Soviet society. Such, however, was neither the intent nor the effect of Stalin's action. He increased his personal power, not the governmental power. When he died, his personal power died with him. The struggle to fill the resulting vacuum was won by Khrushchev, who identified his interests with the interests of the Party organization, rather than by Malenkov, who identified himself with the governmental bureaucracy. Khrushchev's consolidation of power marked the re-emergence and revitalization of the principal organs of the Party. While they acted in very different ways and from different motives, Stalin weakened the Party just as Grant weakened the presidency. Just as a strong presidency is in the American public interest, so also a strong Party is in the Soviet public interest.

In terms of the theory of natural law, governmental actions are legitimate to the extent that they are in accord with the "public philosophy."[53] According to democratic theory, they derive their legitimacy from the extent to which they embody the will of the people. According to the procedural concept, they are legitimate if they represent the outcome of a process of conflict and compromise in which all interested groups have participated. In another sense, however, the legitimacy of governmental actions can be sought in the extent to which they reflect the interests of governmental institutions. In contrast to the theory of representative government, under this concept governmental institutions derive their legitimacy and authority not from the extent to which they represent the interests of the people or of any other group, but from the extent to which they have distinct interests of their own apart from all other groups. Politicians frequently remark that things "look different" after they obtain office than they did when they were competing for office. This difference is a measure of the institutional demands of office. It is precisely this difference in perspective which legitimizes the demands which the officeholder makes on his fellow citizens. The interests of the president, for instance, may coincide partially and temporarily first with those of one group and then with those of another. But the interest of the presidency, as Neustadt has emphasized,[54] coincides with that of no one else. The president's power derives not from his representation of class, group, regional, or popular interests, but rather from the fact that he represents none of these. The presidential perspective is unique to the presidency. Precisely for this reason, it is both a lonely office and a powerful one. Its authority is rooted in its loneliness.

The existence of political institutions (such as the presidency or Presidium) capable of giving substance to public interests distinguishes politically developed societies from undeveloped ones. The "ultimate test of development," as Lucian Pye has said, "is the capacity of a people to establish and maintain large, complex, but flexible organizational forms."[55] The level of organization in much of the world, however, is low. "Except in Europe and America," Banfield notes, "the concerting of behavior in political associations and corporate organizations is a rare and recent thing."[56] The ability to create public organizations and political institutions is in short supply in the world today. It is this ability which, above all else, the Communists offer modernizing countries.

Degeneration and the Corrupt Polity

Most modernizing countries are buying rapid social modernization at the price of political degeneration. This process of decay in political institutions, however, has been neglected or overlooked in much of the literature on modernization. As a result, models and concepts which are hopefully entitled "developing" or "modernizing" are often only partially relevant to the countries to which they are applied. More relevant in many cases would be models of corrupt or degenerating societies, highlighting the decay of political organization and the increasing dominance of disruptive social forces. Who, however, has advanced such a theory of political decay or a model of a corrupt political order which might be useful in analyzing the political processes of the countries that are usually called "developing"? Perhaps the most relevant ideas are the most ancient ones. The evolution of many contemporary new states, once the colonial guardians have departed, has not deviated extensively from the Platonic model. Independence is followed by military coups as the "auxiliaries" take over.[57] Corruption by the oligarchy inflames the envy of rising groups. Conflict between oligarchy and masses erupts into civil strife. Demagogues and street mobs pave the way for the despot. Plato's description of the means by which the despot appeals to the people, isolates and eliminates his enemies, and builds up his personal strength is a far less misleading guide to what has taken place in Ghana and other new states than many things written yesterday.[58]

Plato is one of the few theorists, ancient or contemporary, with a highly explicit theory of political degeneration.[59] The concept of a "corrupt society," however, is a more familiar one in political theory. Typically it refers to a society which lacks law, authority, cohesion, discipline, and consensus, where private interests dominate public ones, where there is an absence of civic obligation and civic duty, where, in short, political institutions are weak and social forces strong. Plato's degenerate states are dominated by various forms of appetite: by force, wealth, numbers, and charisma. "Those constitutions," says Aristotle, "which consider only the personal interest of the rulers are all wrong constitutions, or perversions of the right forms."[60] So also, Machiavelli's concept of the corrupt state, in the words of one commentator, "includes all sorts of license and violence, great inequalities of wealth and power, the destruction of peace and justice, the growth of disorderly ambition, disunion, lawlessness, dishonesty, and contempt for religion."[61] Modern equivalents of the classical corrupt society are Kornhauser's theory of the mass society (where, in the absence of institutions, elites are accessible to masses and masses are available for mobilization by the elite) and Rapoport's concept of the praetorian state where "private ambitions are rarely restrained by a sense of public authority; [and] the role of power (i.e., wealth and force) is maximized."[62] Typical of the corrupt, praetorian, or mass societies is the violent oscillation between extreme democracy and tyranny. "Where the pre-established political authority is highly autocratic," says Kornhauser, "rapid and violent displacement of that authority by a democratic regime is highly favorable to the emergence of extremist mass movements that tend to transform the new democracy in antidemocratic directions."[63] Aristotle and Plato saw despotism emerging out of the extremes of mob rule. Rapoport finds in Gibbon an apt summary of the constitutional rhythms of the praetorian state, which "floats between the extremes of absolute monarchy and wild democracy."[64] Such instability is the hallmark of a society where mobilization has outrun institutionalization.

Strategies of Institutional Development

If decay of political institutions is a widespread phenomenon in the "developing" countries and if a major cause of this decay is the high rate of social mobilization, it behooves us, as social scientists, to call a spade a spade and to incorporate these tendencies into any general model of political change which we employ to understand the politics of these areas. If effective political institutions are necessary for stable and eventually democratic government and if they are also a precondition of sustained economic growth, it behooves us, as policy analysts, to suggest strategies of institutional development. In doing this, we should recognize two general considerations affecting probabilities of success in institution-building.

First, the psychological and cultural characteristics of peoples differ markedly and with them their abilities at developing political institutions. Where age-old patterns of thought and behavior have to be changed, quite obviously the creation of political institutions is a far more difficult task than otherwise. "The Tokugawa Japanese could not, as did the Chinese, put family above government," one expert has observed. "The samurai was expected to be loyal to his official superior first, his family second. In mores generally the primacy of the organization over the person was constantly reiterated."[65] This difference in Japanese and Chinese attitudes toward authority undoubtedly accounts in part for their differences in modernization and development. The Japanese peacefully and smoothly created new political institutions and amalgamated them with old ones. The weakness of traditional Chinese political institutions, on the other hand, led to forty years of revolution and civil war before modern political institutions could be developed and extended throughout Chinese society.

Second, the potentialities for institution-building differ markedly from society to society, but in all societies political organizations can be built. Institutions result from the slow interaction of conscious effort and existing culture. Organizations, however, are the product of conscious, purposeful effort. The forms of this effort may vary from a Meiji Restoration to a Communist Revolution. But in each case a distinct group of identifiable people set about adapting old organizations or building new ones. "Nation-building" has recently become a popular subject, and doubts have been raised about whether nations can be "built."[66] These doubts have a fairly solid basis. Nations are one type of social force, and historically they have emerged over long periods of time. Organization-building, however, differs from nation-building. Political organizations require time for development, but they do not require as much time as national communities. Indeed, most of those who speak of nation-building in such places as tropical Africa see organization-building as the first step in this process. Political parties have to be welded out of tribal groups, the parties create governments, and the governments may, eventually, bring into existence nations. Many of the doubts which people have about the possibilities of nation-building do not apply to organization-building.

Given our hypotheses about the relation of social mobilization to institutionalization, there are two obvious methods of furthering institutional development. First, anything which slows social mobilization presumably creates conditions more favorable to the preservation and strengthening of institutions. Secondly, strategies can be developed and applied directly to the problem of institution-building.

Slowing Mobilization

Social mobilization can be moderated in many ways. Three methods are: to increase the complexity of social structure, to limit or reduce communications in society, and to minimize competition among segments of the political elite.[67]

In general, the more highly stratified a society is and the more complicated its social structure, the more gradual is the process of political mobilization. The divisions between class and class, occupation and occupation, rural and urban, constitute a series of breakwaters which divide the society and permit the political mobilization of one group at a time. On the other hand, a highly homogeneous society, or a society which has only a single horizontal line of division between an oligarchy that has everything and a peasantry that has nothing, or a society which is divided not horizontally but vertically into ethnic and communal groups, has more difficulty moderating the process of mobilization. Thus, mobilization should be slower in India than in the new African states where horizontal divisions are weak and tribal divisions strong, or in those Latin American countries where the middle strata are weak and a small oligarchy confronts a peasant mass. A society with many horizontal divisions gains through the slower entry of social groups into politics. It may, however, also lose something in that political organizations, when they do develop, may form along class and stratum lines and thus lack the autonomy of more broadly based political organizations. Political parties in countries like Chile and Sweden have been largely the spokesmen for distinct classes; caste associations seem destined to play a significant role in Indian politics. The disruptive effects of political organizations identified with social strata may be reduced if other political institutions exist which appeal to loyalties across class or caste lines. In Sweden, loyalty to the monarchy and the Riksdag mitigates the effects of class-based parties, and in India the caste associations must, in general, seek their goals within the much more extensive framework of the Congress Party. In most societies, the social structure must be largely accepted as given. Where it is subject to governmental manipulation and influence, mobilization will be slowed by government policies which enhance the complexity of social stratification.

The communications network of a society is undoubtedly much more subject to governmental influence. Rapid gains in some of the most desired areas of modernization—such as mass media exposure, literacy, and education—may have to be purchased at the price of severe losses in political stability. This is not to argue that political institutionalization as a value should take precedence over all others: if this were the case, modernization would never be good. It is simply to argue that governments must balance the values won through rapid increases in communications against the values jeopardized by losses in political stability. Thus, governmental policies may be wisely directed to reducing the number of university graduates, particularly those with skills which are not in demand in the society. Students and unemployed university graduates have been a concern common to the nationalistic military regime in South Korea, the socialist military regime in Burma, and the traditional military regime in Thailand. The efforts by General Ne Win in Burma to cut back the number of university graduates may well be imitated by other governments facing similar challenges. Much has been made of the problems caused by the extension of the suffrage to large numbers of illiterates. But limited political participation by illiterates may well, as in India, be less dangerous to political institutions than participation by literates. The latter typically have higher aspirations and make more demands on government. Political participation by illiterates, moreover, is more likely to remain limited, whereas participation by literates is much more likely to snowball with potentially disastrous effects on political stability. A governing elite may also affect the intensity of communications and the rate of political mobilization by its policies on economic development. Large, isolated factories, as Kornhauser has shown, are more likely to give rise to extremist movements than smaller plants more closely integrated into the surrounding community.[68] Self-interest in political survival may lead governing elites to decrease the priority of rapid economic change.

The uncontrolled mobilization of people into politics is also slowed by minimizing the competition among political elites. Hence mobilization is likely to have less disturbing effects on political institutions in one-party systems than in two-party or multiparty systems. In many new states and modernizing countries, a vast gap exists between the modernized elite and the tradition-oriented mass. If the elite divides against itself, its factions appeal to the masses for support. This produces rapid mobilization of the masses into politics at the same time that it destroys whatever consensus previously existed among the politically active on the need for modernization. Mobilization frequently means the mobilization of tradition; modern politics become the vehicle of traditional purposes. In Burma during the first part of this century, the "general pattern was one in which the modernizers first fell out among themselves whenever they were confronted with demanding choices of policy, and then tended to seek support from among the more traditional elements, which in time gained the ascendency."[69] In Turkey a rough balance between the mobilization of people into politics and the development of political institutions existed so long as the Republican People's Party retained a political monopoly. The conscious decision to permit an opposition party, however, broadened the scope of political competition beyond the urban, Westernized elite. The Democratic Party mobilized the peasants into politics, strengthened the forces of traditionalism, and broke the previous consensus. This led the party leaders to attempt to maintain themselves in power through semilegal means and to induce the army to join them in suppressing the Republican opposition. The army, however, was committed to modernization and seized power in a coup d'etat, dissolving the Democratic Party and executing many of its top leaders. In due course, the military withdrew from direct conduct of the government, and democratic elections led to a multiparty system in which no party has a clear majority. Thus, from a relatively stable one-party system, Turkey passed through a brief two-party era to military rule and a multiparty system: the familiar syndrome of states where mobilization has outrun institutionalization. In the process, not only were political institutions weakened, but the traditional-minded were brought into politics in such a way as to create obstacles to the achievement of many modernizing goals.

Creating Institutions

"Dans la naissance des sociétés ce sont les chefs des républiques qui font l'institution; et c'est ensuite l'institution qui forme les chefs des républiques," said Montesquieu.[70] But in the contemporary world, political leaders prefer modernization to institution-building, and no matter who leads modernization, the process itself generates conflicting demands and inducements which obstruct the growth of political institutions. Where modernization is undertaken by traditional leaders working through traditional political institutions, the efforts of the traditional leaders to reform can unleash and stimulate social forces which threaten the traditional political institutions. Traditional leaders can modernize and reform their realms, but, where substantial social elements oppose reform, they have yet to demonstrate they can put through reforms without undermining the institutions through which they are working. The problem is: How can the traditional political institutions be adapted to accommodate the social forces unleashed by modernization? Historically, except for Japan, traditional political institutions have been adapted to the modern world only where a high degree of political centralization was not required for modernization and where traditional (i.e., feudal) representative institutions retained their vitality (as in Great Britain and Sweden). If modernization requires the centralization of power in a "reform monarch" or "revolutionary emperor," it means the weakening or destruction of whatever traditional representative institutions may exist and thus complicates still further the assimilation of those social forces created by modernization. The concentration of power also makes the traditional regime (like the eighteenth-century French monarchy) more vulnerable to forcible overthrow. *The vulnerability of a traditional regime to revolution varies directly with the capability of the regime for modernization.* For traditional rulers, the imperatives of modernization conflict with the imperatives of institution-building.

If the traditional political institutions are weak, or if they have been displaced and suppressed during periods of colonial rule, adaptation is impossible. In societies which have undergone colonial rule, incubation can serve as a substitute for adaptation. Unfortunately, the opportunity for incubation was missed in most colonial societies, with a few prominent exceptions such as India and the Philippines. Incubation requires a colonial administration which is willing to permit and *to contend with* a nationalist movement for many years, thus furnishing the time, the struggle, and the slowly increasing responsibility which are the ingredients of institution-building. In general, however, colonial powers tend to postpone incubation for as long as possible and then, when they see independence as inevitable, to bring it about as quickly as possible. Consequently, most of the states which became independent in the 1950s and 1960s had little opportunity to incubate political institutions while still under colonial tutelage.

Where traditional political institutions are weak, or collapse, or are overthrown, authority frequently comes to rest with charismatic leaders who attempt to bridge the gap between tradition and modernity by a highly personal appeal. To the extent that these leaders are able to concentrate power in themselves, it might be supposed that they would be in a position to push institutional development and to perform the role of "Great Legislator" or "Founding Father." The reform of corrupt states or the creation of new ones, Machiavelli argued, must be the work of one man alone. A conflict exists, however, between the interests of the individual and the interests of institutionalization. Institutionalization of power means the limitation of power which might otherwise be wielded personally and arbitrarily. The would-be institution-builder needs personal power to create institutions but he cannot create institutions without relinquishing personal power. Resolving this dilemma is not easy. It can be done only by leaders who combine rare political skill and rare devotion to purpose. It was done by Mustafa Kemal who, for almost two decades, managed to maintain his own personal power, to push through major modernizing reforms, and to create a political institution to carry on the government after his death. Ataturk has been a conscious model for many contemporary modernizing leaders, but few, if any, seem likely to duplicate his achievement.

The military junta or military dictatorship is another type of regime common in modernizing countries. It too confronts a distinct set of problems in the conflict between its own impulses to modernization and the needs of institution-building. The military officers who seize power in a modernizing country frequently do so in reaction to the "chaos," "stalemate," "corruption," and "reactionary" character of the civilian regimes which preceded them. The officers are usually passionately devoted to measures of social reform, particularly those which benefit the peasantry (whose interests have frequently been overlooked by the anterior civilian regime). A rationalistic approach to societal problems often makes the officers modernizers par excellence. At the same time, however, they are frequently indifferent or hostile to the needs of political institution-building. The military typically assert that they have taken over the government only temporarily until conditions can be "cleaned up" and authority restored to a purified civilian regime. The officers thus confront an organizational dilemma. They can eliminate or exclude from politics individual civilian politicians, but they are ill-prepared to make fundamental changes in political processes and institutions. If they turn back power to the civilians, the same conditions to which they originally objected tend to reappear (Burma). If they attempt to restore civilian government and to continue in power as a civilian political group (Turkey, South Korea), they open themselves to these same corrupting influences and may pave the way for a second military takeover by a younger generation of colonels who purge the civilianized generals, just as the generals had earlier purged the civilians. Finally, if the military leaders retain power indefinitely, they need to create authoritative political organizations which legitimize and institutionalize their power. Concern with their own personal authority and unfamiliarity with the needs of political institution-building create problems in the fulfillment of this task. It is still too early to say for certain what sort of authoritative political institutions, if any, will be produced by regimes led by military officers such as Nasser and Ayub Khan.

The Primacy of Party

Charismatic leaders and military chiefs have thus had little success in building modern political institutions. The reason lies in the nature of modern politics. In the absence of traditional political institutions, the only modern organization which can become a source of authority and which can be effectively institutionalized is the political party. *The importance of the political party in providing legitimacy and stability in a modernizing political system varies inversely with the institutional inheritance of the system from traditional society.* Traditional systems do not have political parties. Unlike bureaucracy, the party is a distinctly modern form of political organization. Where traditional political institutions (such as monarchies and feudal parliaments) are carried over into the modern era, parties play secondary, supplementary roles in the political system. The other institutions are the primary source of continuity and legitimacy. Parties typically originate within the legislatures and then gradually extend themselves into society. They adapt themselves to the existing framework of the political system and typically reflect in their own operations the organizational and procedural principles embodied in that system. They broaden participation in the traditional institutions, thus adapting those institutions to the requirements of the modern polity. They help make the traditional institutions legitimate in terms of popular sovereignty, but they are not themselves a source of legitimacy. Their own legitimacy derives from the contributions they make to the political system.

Where traditional political institutions collapse or are weak or nonexistent, the role of the party is entirely different from what it is in those polities with institutional continuity. In such situations, strong party organization is the only long-run alternative to the instability of a corrupt or praetorian or mass society. The party is not just a supplementary organization; it is instead the source of legitimacy and authority. In the absence of traditional sources of legitimacy, legitimacy is sought in ideology, charisma, popular sovereignty. To be lasting, each of these principles of legitimacy must be embodied in a party. Instead of the party reflecting the state, the state becomes the creation of the party and the instrument of the party. The actions of government are legitimate to the extent that they reflect the will of the party. The party is the source of legitimacy because it is the institutional embodiment of national sovereignty, the popular will, or the dictatorship of the proletariat.

Where traditional political institutions are weak or nonexistent, the prerequisite of stability is at least one highly institutionalized political party. States with one such party are markedly more stable than states which lack such a party. States with no parties or many weak parties are the least stable. Where traditional political institutions are smashed by revolution, postrevolutionary order depends on the emergence of one strong party: witness the otherwise very different histories of the Chinese, Mexican, Russian, and Turkish revolutions. Where new states emerge from colonialism with one strong party, the problem is to maintain the strength of that party. In many African countries the nationalist party was the single important modern organization to exist before independence. The party "was generally well organized. The conditions of the political struggle and the dedication of the top elite to the party as the prime instrument of political change led the elite to give the major portion of their energies and resources to building a solid, responsive organization capable of disciplined action in response to directives from the top and able to ferret out and exploit feelings of dissatisfaction among the masses for political ends."[71] After independence, however, the dominant political party is often weakened by the many competing demands on organizational resources. A marked dispersion of resources means a decline in the overall level of political institutionalization. "Talents that once were available for the crucial work of party organization," one observer has warned, "may now be preoccupied with running a ministry or government bureau. . . . Unless new sources of loyal organizational and administrative talents can be found immediately, the party's organization—and, therefore, the major link between the regime and the masses—is likely to be weakened."[72]

The need for concentration applies not only to the allocation of resources among types of organizations but also to the scope of organization. In many modernizing countries, the political leaders attempt too much too fast; they try to build mass organizations when they should concentrate on elite organizations. Organizations do not have to be large to be effective and to play a crucial role in the political process: the Bolshevik Party in 1917 is one example; the Indian Civil Service (which numbered only 1,157 men at independence) is another. Overextension of one's resources in organization-building is as dangerous as overextension of one's troops in a military campaign. (The strategic hamlet program in South Vietnam is an example of both.) Concentration is a key principle of politics as well as strategy. The pressures for broad organizational support, however, seem to push towards the all-inclusive organization. In his efforts to create a political structure to bolster his military regime in Egypt, for instance, Nasser first created the Liberation Rally in 1953, which soon came to have from 5 to 6 million members. The organization was simply too big to be effective and to achieve its purpose. After the adoption of a new constitution in 1956, the Liberation Rally was replaced by the National Union, which was designed to be the school of the nation and also to be universal in membership (except for reactionaries). Again the organization was too broad to be effective. Hence in 1962, after the break with Syria, a new organization, the Arab Socialist Union, was organized with the advice of organizational and ideological experts from Yugoslavia. It was designed to be a more exclusive, more tightly organized body, its membership limited to 10 percent of the population. Inevitably, however, it also mushroomed in size, and after two years it had 5 million members. In a fourth effort, early in 1964 President Nasser reportedly formed still another group limited to only 4,000 members and called the "Government Party," which would form the core of the Arab Socialist Union. The new organization was to be designed by Nasser "to enforce a peaceful transfer of power and a continuation of his policies if anything happens to him."[73] Whether this organization, unlike its predecessors, becomes an institution remains to be seen. Its likelihood of success depends upon its limitation in size.

American social scientists have devoted much attention to the competitiveness of political systems, devising various ways of measuring that competitiveness and classifying systems according to their degree of competitiveness.[74] The more parties which exist within a system, presumably the more competitive it is. Yet the proliferation of parties usually means the dispersion of organization and leadership talents and the existence of a large number of weak parties. If sufficient resources are available to support more than one well-organized party, this is all to the good. But most modernizing countries will be well off if they can create just one strong party organization. *In modernizing systems, party institutionalization usually varies inversely with party com-*

petitiveness. Modernizing states with multiparty systems are much more unstable and prone to military intervention than modernizing states with one party, with one dominant party, or with two parties. The most unstable systems and those most prone to military intervention are the multiparty systems and the no-party systems. The weak institutionalization of parties in the multiparty system makes that system extremely fragile. The step from many parties to no parties and from no parties to many parties is an easy one. In their institutional weakness, the no-party system and the multiparty system closely resemble each other.

Political Development and American Policy

The Bolshevik concept of the political party is directly relevant to modernizing countries. It provides a conscious and explicit answer to the problem of mobilization versus institutionalization. The Communists actively attempt to expand political participication. At the same time they are the most energetic and intense contemporary students of de Tocqueville's "art of associating together." Their specialty is organization, their goal the mobilization of the masses into their organizations. For them mobilization and organization go hand in hand. "There are only

TABLE 3: *Distribution of Coups and Coup Attempts in Modernizing Countries Since Independence*

Type of Political System	*Number of Countries*	*Countries with Coups*	
		number	*percent*
Communist	3	0	0
One-party	18	2	11
One-party dominant	12	3	25
Two-party	11	5	45
Multiparty	22	15	68
No effective parties	17	14	83

Source: Figures are somewhat revised and adapted from the similar table in Fred R. von der Mehden, *Politics of the Developing Nations* (Englewood Cliffs, N.J., 1964), p. 65.

two kinds of political tasks," a leading Chinese Communist theorist has said: "one is the task of propaganda and education, the other is the task of organization."[75] The party is initially a highly select group of those who have achieved the proper degree of revolutionary consciousness. It expands gradually as it is able to win the support and participation of others. Peripheral organizations and front groups provide an organizational ladder for the gradual mobilization and indoctrination of those who in due course become full-fledged party members. If the political struggle takes the form of revolutionary war, mobilization occurs on a gradual territorial basis as village after village shifts in status from hostile control to contested area to guerrilla area to base area. The theory is selective mobilization; the political involvement of masses who have not reached the proper level of revolutionary consciousness can only benefit reaction. The "opportunist" Menshevik, Lenin warned, "strives to proceed from the bottom upward, and, therefore, wherever possible and as far as possible, upholds autonomism and 'democracy'. . . ." The Bolshevik, on the other hand, "strives to proceed from the top downward, and upholds an extension of the rights and powers of the center in relation to the parts."[76]

Communist doctrine thus recognizes the need to balance mobilization and organization and stresses the party as the key to political stability. The American approach, on the other hand, tends to ignore the requirements of political organization and to deprecate the importance of party. American attitudes are rooted in the secondary, instrumental role of party in the American constitutional system. In addition, American distaste for politics leads to an emphasis on the output aspects of the political system. The stress, as Lucian Pye has pointed out, has been on the efficient administrator rather than the wily politician.[77] Aid missions advise governments on administrative organization and economic planning, but seldom do they advise political leaders on how to create a strong party. To meet the problems of interest aggregation, Americans have resorted to a variety of alternatives to political organization. Some Americans have urged that the military, as the strongest organization in many modernizing countries, should assume a major role in the responsibilities of government.[78] At times American policy has relied on individual political leaders, such as Magsaysay, Diem, or Ayub Khan. Alternatively, the American government has stressed adherence to particular structural forms, such as free elections.

All of these approaches are doomed to failure. Neither military juntas nor charismatic personalities nor free elections can be a long-term substitute for effective political organization. Charismatic leaders are reluctant to substitute party control for personal control. Military officers are usually even more explicitly antiparty. They contrast the venal party politics of the civilians with the honest devotion to the nation of the military. Military coups and military juntas may spur modernization, but they cannot produce a stable political order. Instead of relying on the military, American policy should be directed to the creation within modernizing countries of at least one strong non-Communist political party. If such a party already exists and is in a dominant position, support of that party should be the keystone of policy. Where political life is fragmented and many small parties exist, American backing should go to the strongest of the parties whose goals are compatible with ours. If it is a choice between a party and a personality, choose the party: better the Baath than Nasser. Where no parties exist and the government (whether traditional, military, or charismatic) is reasonably cooperative with the United States, American military, economic, and technical assistance should be conditioned upon the government's making efforts to develop a strong supporting party organization.

Several years ago Guy Pauker warned that "What is most urgently needed in Southeast Asia today is organizational strength."[79] Organizational strength is also the most urgent need in southern Asia, the Middle East, Africa, and Latin America. Unless that need is met with American support, the alternatives in those areas remain a corrupt political system or a Communist one.

NOTES

1. *Democracy in America,* Phillips Bradley, ed. (New York, 1955), pp. 11, 118.
2. Lucian W. Pye, ed., *Communications and Political Development* (Princeton, 1963), p. 16.
3. Robert E. Ward and Dankwart A. Rustow, eds., *Political Modernization in Japan and Turkey* (Princeton, 1964), pp. 6–7; Rupert Emerson, *Political Modernization: The Single-Party System* (Denver, 1963), pp. 7–8; Pye, ed., *Communications and Political Development,* pp. 17–18; S. N. Eisenstadt, "Bureaucracy and Political Development," in Joseph LaPalombara, ed., *Bureaucracy and Political Development* (Princeton, 1963), p. 99.
4. James S. Coleman, in Gabriel A. Almond and Coleman, eds., *The Politics of the Developing Areas* (Princeton, 1960), p. 532; Fred W. Riggs, "Bureaucracy and Political Development: A Paradoxical View," in LaPalombara, *ed., Bureaucracy and Political Development,* p. 122; Eisenstadt, in *ibid.,* p. 99; Ward and Rustow, eds., *Political Modernization,* p. 7.
5. See, e.g., Gabriel A. Almond, "Political Systems and Political Change," *American Behavioral Scientist* 6 (June 1963), 3–10; Ward and Rustow, eds., *Political Modernization,* p. 7.
6. Coleman, in Almond and Coleman, eds., *Politics of Developing Areas,* p. 533.
7. Frederick W. Frey, "Political Development, Power, and Communications in Turkey," in Pye, ed., *Communications and Political Development,* p. 301.
8. Karl W. Deutsch, "Social Mobilization and Political Development," *American Political Science Review* 55 (September 1961), 493 ff.
9. Daniel Lerner, *The Passing of Traditional Society* (Glencoe, 1958), pp. 48–50.
10. Gabriel A. Almond and Sidney Verba, *The Civic Culture* (Princeton, 1963), p. 4.
11. Dankwart A. Rustow, "The Vanishing Dream of Stability," *AID Digest* (August 1962), p. 13; Riggs, in LaPalombara, ed., *Bureaucracy and Political Development,* p. 139.
12. Lerner, *Passing of Traditional Society,* chap. 2.
13. For the *reductio ad absurdum,* see Majid Khadduri, *Modern Libya: A Study in Political Development* (Baltimore, 1963), and J. Clagett Taylor, *The Political Development of Tanganyika* (Stanford, Cal., 1963). In the titles and content of both, "political development" has no analytical meaning. It is simply a synonym (euphemism?) for "political history." Both books are good history, but they are not social science.
14. See, e.g., Milton J. Esman, "The Politics of Development Administration," in John D. Montgomery and William Siffin, eds., *Politics,*

Administration, and Change: Approaches to Development (New York, 1965). Esman bases his analysis on the assumption that the political leaders of modernizing societies are motivated by the goals of nation-building and social-economic progress and not by desire for personal power, wealth, status, or the territorial expansion of their countries. This assumption has about the same degree of truth and usefulness in explaining politics in the contemporary "developing" areas as the assumption that Stalin's policies were devoted to building communism has to the explanation of Soviet politics in the 1930s.

15. On the "erosion of democracy" and political instability, see Rupert Emerson, *From Empire to Nation* (Cambridge, Mass., 1960), chap. 15; and Michael Brecher, *The New States of Asia* (London, 1963), chap. 2.
16. Almond, "Political Systems and Political Change," p. 6.
17. The concept of institutionalization has, of course, been used by other writers concerned with political development—most notably, S. N. Eisenstadt. His definition, however, differs significantly from my approach here. See, in particular, his "Initial Institutional Patterns of Political Modernisation," *Civilisations* 12, no. 4 (1962), 461–472, and 13, no. 1 (1963), 15–26; "Institutionalization and Change," *American Sociological Review* 29 (April 1964), 235–247; "Social Change, Differentiation, and Evolution," *ibid.*, 29 (June 1964), 375–386.
18. Ashoka Mehta, in Raymond Aron, ed., *World Technology and Human Destiny* (Ann Arbor, Mich., 1963), p. 133.
19. See the very useful discussion in Philip Selznick's small classic, *Leadership in Administration* (New York, 1957), pp. 5 ff.
20. See Mayer N. Zald and Patricia Denton, "From Evangelism to General Service: The Transformation of the YMCA," *Administrative Science Quarterly* 8 (September 1963), 214 ff.
21. Joseph R. Gusfield, "Social Structure and Moral Reform: A Study of the Woman's Christian Temperance Union," *American Journal of Sociology* 61 (November 1955), 232; and Gusfield, "The Problem of Generations in an Organizational Structure," *Social Forces*, 35 (May 1957), 323 ff.
22. Sheldon L. Messinger, "Organizational Transformation: A Case Study of a Declining Social Movement," *American Sociological Review* 20 (February 1955), 10; italics in original.
23. David L. Sills, *The Volunteers* (Glencoe, 1957), p. 266. Chap. 9 of this book is an excellent discussion of organizational goal replacement with reference to the YMCA, WCTU, Townsend Movement, Red Cross, and other case studies.
24. Sigmund Neumann, "Toward a Comparative Study of Political Parties," in Neumann, ed., *Modern Political Parties* (Chicago, 1956), pp. 403–405.
25. *Politics,* trans. Ernest Barker (London, 1946), p. 254.
26. *Reflections on the Revolution in France,* Gateway ed. (Chicago, 1955), p. 37.
27. *Politics,* 60, 206.
28. *Reflections on the Revolution in France,* 92.

29. See Samuel P. Huntington, "Patterns of Violence in World Politics," in Huntington, ed., *Changing Patterns of Military Politics* (New York, 1962), pp. 44–47.
30. See, e.g., Herbert McCloskey, "Consensus and Ideology in American Politics," *American Political Science Review* 28 (June 1964), 361 ff.; Samuel Stouffer, *Communism, Conformity, and Civil Liberties* (New York, 1955), *passim.*
31. Arnold J. Toynbee, *A Study of History,* abridgement of vols. 1–6 by D. C. Somervell (New York, 1947), pp. 176–177.
32. David C. Rapoport, "A Comparative Theory of Military and Political Types," in Huntington, ed., *Changing Patterns of Military Politics,* p. 79.
33. Harry Holbert Turney-High, *Primitive War* (Columbia, S.C., 1949), pp. 235–236.
34. William Kornhauser, *The Politics of Mass Society* (Glencoe, 1959), p. 145.
35. *Ibid.,* p. 146.
36. See Mancur Olson, Jr., "Rapid Growth as a Destabilizing Force," *Journal of Economic History* 27 (December 1963), 529–552; and Bert F. Hoselitz and Myron Weiner, "Economic Development and Political Stability in India," *Dissent* 8 (Spring 1961), 172–179.
37. See Daniel Lerner, "Toward a Communication Theory of Modernization," in Pye, ed., *Communications and Political Development,* pp. 330 ff.
38. Cf. Deutsch, "Social Mobilization and Political Development," p. 496.
39. Daniel Lerner, "The Transformation of Institutions," mimeographed, p. 19.
40. John J. Johnson, *The Military and Society in Latin America* (Stanford, Cal., 1964), pp. 98–99.
41. Lucian W. Pye, *Politics, Personality, and Nation Building* (New Haven, 1962), pp. 4–5.
42. See, in general, Ronald E. Wraith and Edgar Simpkins, *Corruption in Developing Countries* (London, 1963).
43. These figures are calculated from the data in the Appendix of Fred R. von der Mehden, *Politics of the Developing Nations* (Englewood Cliffs, N.J., 1964).
44. Computed from figures in R. W. Fitzgibbon, "Armies and Politics in Latin America," paper, 7th Round Table, International Political Science Association, Opatija, Yugoslavia, September 1959, pp. 8–9.
45. Thomas B. Macaulay, letter to Henry S. Randall, Courtlandt Village, New York, May 23, 1857, printed in "What Did Macaulay Say About America?" *Bulletin of the New York Public Library* 29 (July 1925), 477–479.
46. Simon Bolivar, quoted in K. H. Silvert, ed., *Expectant Peoples: Nationalism and Development* (New York, 1963), p. 347.
47. *El Dia,* Quito, November 27, 1943, quoted in Bryce Wood, *The Making of the Good Neighbor Policy* (New York, 1961), p. 318.
48. Ralph Braibanti, "Public Bureaucracy and Judiciary in Pakistan," in LaPalombara, ed., *Bureaucracy and Political Development,* p. 373.

49. Johnson, *Military and Society*, p. 143.
50. See, in general, Glendon Schubert, *The Public Interest* (Glencoe, 1960); Carl J. Friedrich, ed., *The Public Interest*, Nomos V (New York, 1962); Douglas Price, "Theories of the Public Interest," in Lynton K. Caldwell, ed., *Politics and Public Affairs* (Bloomington, Ind., 1962), pp. 141–160.
51. Carl J. Friedrich, *Man and His Government* (New York, 1963), p. 150; italics in original.
52. *Politics*, 267.
53. See Walter Lippmann, *The Public Philosophy* (Boston, 1955), especially p. 42, for his definition of the public interest as "what men would choose if they saw clearly, thought rationally, acted disinterestedly and benevolently."
54. See Richard E. Neustadt, *Presidential Power* (New York, 1960), *passim*, but especially pp. 33–37, 150–151.
55. Pye, *Politics, Personality, and Nation Building*, p. 51.
56. Edward C. Banfield, *The Moral Basis of a Backward Society* (Glencoe, Ill., 1958), pp. 7–9, 15 ff.
57. For comments on the short time lag between independence and the first coup, see Dankwart A. Rustow, "The Military in Middle Eastern Society and Politics," in Sydney N. Fisher, ed., *The Military in the Middle East: Problems in Society and Government* (Columbus, Ohio, 1963), p. 10.
58. See, in general, *The Republic*, Book 8, and especially the description of the despotic regime, trans. Cornford (New York, 1945), pp. 291–293.
59. Perhaps the closest contemporary model comes not from a social scientist but from a novelist, William Golding. The schoolboys (newly independent elites) of *The Lord of the Flies* initially attempt to imitate the behavior patterns of adults (former Western rulers). Discipline and consensus, however, disintegrate. A demagogic military leader and his followers gain or coerce the support of a majority. The symbol of authority (the conch) is broken. The voices of responsibility (Ralph) and reason (Piggy) are deserted and harassed, and reason is destroyed. In the end, the naval officer (British Marine Commandos) arrives just in time to save Ralph (Nyerere) from the "hunters" (mutinous troops).
60. *Politics*, 112.
61. George H. Sabine, *A History of Political Thought*, rev. ed. (New York, 1950), p. 343.
62. Kornhauser, *Politics of Mass Society*, *passim;* David C. Rapoport, "Praetorianism: Government without Consensus" (Ph.D. dissertation, University of California, Berkeley, 1959); and Rapoport in Huntington, ed., *Changing Patterns of Military Politics*, p. 72, where the quotation occurs.
63. Kornhauser, *Politics of Mass Society*, p. 125.
64. Edward Gibbon, *The Decline and Fall of the Roman Empire* (New York, 1899), Vol. 1, p. 235, quoted by Rapoport in Huntington, ed., *Changing Patterns of Military Politics*, p. 98.
65. John Whitney Hall, "The Nature of Traditional Society: Japan," in Ward and Rustow, eds., *Political Modernization in Japan and Turkey*, p. 19.

66. See Karl W. Deutsch and William J. Foltz, eds., *Nation-Building* (New York, 1963), *passim,* but especially the contributions of Joseph R. Strayer and Carl J. Friedrich.
67. These are not, of course, the only ways of slowing mobilization. Myron Weiner, for instance, has suggested that one practical method is "localization": channeling political interests and activity away from the great issues of national politics to the more immediate and concrete problems of the village and community. This is certainly one motive behind both community development programs and "basic democracies."
68. Kornhauser, *Politics of Mass Society,* pp. 150–158.
69. Pye, *Politics, Personality, and Nation Building,* p. 114.
70. Charles de Secondat, Baron Montesquieu, *Considérations sur les causes de la grandeur des romains et de leur décadence,* in *Oeuvres* (Paris, 1828), Vol. 1, pp. 119–120.
71. William J. Foltz, "Building the Newest Nations: Short-Run Strategies and Long-Run Problems," in Deutsch and Foltz, eds., *Nation-Building,* p. 121.
72. *Ibid.,* pp. 123–124.
73. *Washington Post,* February 9, 1964, p. A–17.
74. See James S. Coleman, in Almond and Coleman, eds., *Politics of the Developing Areas,* Conclusion; Philips Cutright, "National Political Development: Its Measurement and Social Correlates," in Nelson W. Polsby, Robert A. Dentler, and Paul A. Smith, eds., *Politics and Social Life* (Boston, 1963), pp. 569–582; von der Mehden, *Politics of the Developing Nations,* pp. 54–64.
75. Ai Ssu-chi, quoted in Frederick T. C. Yu, "Communications and Politics in Communist China," in Pye, ed., *Communications and Political Development,* pp. 261–262.
76. V. I. Lenin, *One Step Forward, Two Steps Back (The Crisis in Our Party),* in *Collected Works,* (trans. Fineberg and Jochel) (London, 1961), pp. 396–397.
77. Pye, *Politics, Personality, and Nation Building,* pp. 297–301; and Pye, "The Policy Implications of Social Change in Non-Western Societies," M.I.T. Center for International Studies (Cambridge, 1957), pp. 69–80 (mimeographed).
78. Guy J. Pauker, "Southeast Asia as a Problem Area in the Next Decade," *World Politics* 11 (April 1959), 325–345; Lucian W. Pye, "Armies in the Process of Political Modernization," in John J. Johnson, ed., *The Role of the Military in Underdeveloped Countries* (Princeton, 1962), pp. 69–90.
79. Pauker, "Southeast Asia as a Problem Area in the Next Decade," p. 343.

7 *The African Military and Political Development*

CLAUDE E. WELCH, JR.

Once civilian governments have been ousted, the leaders of military intervention seek to justify their seizure of control. To "prove" that installation of an army regime was necessary is a first order of business. But the newly installed rulers must go a step further and proclaim their goals for the country. Possibly the most important theme sounded by the military leaders has been "national reconstruction." Though wording varies, the import remains identical: Politicians have failed to resolve the fundamental economic, political, and social problems confronting the state; only a transitional period of military rule can purge the political system of its inadequacies.

Presented at the annual meeting of the African Studies Association, New York City, November 2, 1967, and published here by permission of the author. Preparation of this paper was aided by a fellowship from the Research Foundation of the State University of New York. Portions of this chapter have been published, in slightly revised form, in *Soldier and State in Africa,* edited by Claude E. Welch, Jr. (Evanston, Ill.: Northwestern University Press, 1970).

Rhetoric must not be confused with probability of action, however. The promises of renovation and rebuilding may prove hollow. The fundamental question remains: Can a military-based government cope any more successfully with the difficulties civilian regimes encountered? Are some of these problems susceptible to solution by means congenial to the governing military junta, in ways that escaped the preceding civilian regime? And, pursuing a different line of questioning, will the armed forces carry through on the oft-repeated promise of a rapid return to the barracks?

Responses to these questions must be hedged with caution. Military intervention in postcolonial Africa is too recent a phenomenon to permit generalizations based on extensive historical evidence. Analogies between patterns of intervention in Africa and elsewhere in the *tiers monde* should not be extrapolated into sweeping observations. Our knowledge of social processes in contemporary Africa remains sketchy. These caveats, however, should not preclude tentative answers to such questions; clearly, the African military has become a major determinant of political activity.

The problem I shall confront in this paper is vast, both in its implications and its significance. Can the African military, in attempting to carry out "national reconstruction," maintain or create institutions capable of resolving major political difficulties? Or, in slightly different terms, in what ways might army rule contribute to political development? Our focus accordingly, will be upon the significant—and speculative—area of purposeful political change.

The constituent elements of political development have been subject to heated debate among scholars—and, as a natural outgrowth of dispute, have generated extensive commentary.[1] Three major aspects recur in most definitions: (1) an increased centralization of power in the state, coupled with the weakening of traditional sources of authority; (2) the differentiation and specialization of political institutions; (3) increased popular participation in politics and greater identification with the political system as a whole.[2] I shall analyze these aspects in general terms, to generate tentative hypotheses about the impact of military, in the first section. The second section of the paper will briefly touch upon twentieth-century political change in Turkey, features of which may be relevant to contemporary Africa. In the third section, I shall focus upon emerging patterns of military rule south of the Sahara, and I shall reformulate the tentative hypotheses in light of available evidence about political development and army rule.

I. The Military and Political Development: Theoretical Connections

Centralization of Authority

As organizations, armed forces are characterized by centralization, discipline, hierarchy, communication, and esprit de corps.[3] To function effectively, they require a clearly defined chain of command with adequate communications to ensure that orders are carried out and with means of disciplinary control. Effective military organization, almost by definition, demands a high degree of centralization.

Many scholars have equated "modern" government with centralized and highly organized government. Samuel P. Huntington has argued in the following manner: "Political modernization . . . involves the rationalization of authority: the replacement of a large number of traditional, religious, familial, and ethnic political authorities by a single, secular, national political authority. . . . It means national integration and the centralization or accumulation of power in recognized national law-making institutions."[4] Centralization of power, it would appear, accords well with patterns of organization familiar to the military. Hence, one can hypothesize that a coup d'etat will likely result in greater control exercised from the center.

Perhaps the key word in Huntington's definition is authority.

Without entering into the disputations that have surrounded this word,[5] authority may be defined as the acceptance of certain individuals or offices as possessing a legitimate right to leadership. The exercise of authority presupposes "a community of opinions, values, and beliefs, as well as of interests and needs."[6] Applied to the context of military intervention in politics, the question of authority centers on the act of seizing control: Is this regarded as usurpation, or as a rightful act? In other words, by their ousting of the civilian government, do the armed forces become the legitimate wielders of authority? Can the military develop the "community of opinions, values, and beliefs, as well as of interests and needs" that constitute authority?

For a tentative answer, let us turn briefly to Hobbe's analogy between politics and a game of cards. In both, Hobbes suggested, the participants must agree upon what constitutes legitimate power—what is authority; which cards are trumps. If no card is accepted, then clubs—force of arms—become trumps. This leads to our second hypothesis. Where confusion prevails over the source of authority, the possibility of military intervention increases significantly; should intervention occur, the leaders of the coup will attempt to convert their power into authority.[7] Their prospects for success, I shall suggest subsequently, depend upon developing political institutions with a capacity for effective change.

Weakening of traditional sources of authority does not necessarily result from centralization of government functions. To replace traditional, religious, familial, and ethnic political authorities by a single, secular, national political authority obviously requires time, favorable conditions, communications based upon similar values, a growth in mutual confidence.[8] In the absence of favorable conditions, attempts at centralization may touch off major countervailing forces. Tensions may be exacerbated; would-be centralization can bring disintegration. The experiences of Nigeria under the Ironsi government seem to justify this hypothesis.[9] Accordingly, I would be hesitant to agree fully with Guy Pauker, who (writing about Southeast Asia) asserts that the officer corps can become "temporary kernels of national integration, around which the other constructive forces of the various societies could rally, during a short period of breakthrough from present stagnation into a genuine developmental takeoff."[10]

Differentiation and Specialization

In a traditional setting, the many functions carried out within a society may be "fused," in Rigg's phrase.[11] A modern setting, on the other hand, is characterized by differentiation among these functions and the development of particular structures (institutions) for their accomplishment. Clearly, the armed services epitomize such differentiation of function and specialization of structure. A modern army is not a temporary band of marauders or a disorganized militia. Its raison d'etre is the rational utilization of violence; its organization reflects the application of modern techniques.[12]

The high degree of specialization within the armed services contrasts not only with the "fused" traditional social setting but also with other "modern" groups. Few political parties in Africa, for example, can match the centralization, discipline, hierarchy, esprit de corps, and speed of communications manifested by even the smallest professional armies on the continent. Qualitatively, the organizational strength of the military sets it apart from other groups. Command of weaponry is not the sole distinguishing factor of African armies vis-à-vis civilian organizations: the very organizational characteristics of armies are distinctive.

Many commentators on the role of the military in developing countries have focused upon the impact of army training upon both officers and recruits. It stands to reason that extended military service, under certain conditions, will weaken an individual's primary identification with his village and ethnic group. An African soldier must likely learn a new language, serve in areas hundreds of miles from his home, and associate with men of different ethnic backgrounds. He may acquire technological skills unknown to the average villager. Specialized training courses for officers (and particularly those taken outside Africa) further remove them from the traditional setting. Lucian W. Pye argues in this fashion:

> the good soldier is also to some degree a modernized man. Thus it is that the armies in the newly emergent countries come to play key roles in the process by which traditional ways give way to more Westernized ideas and practices . . . the acculturative process in the army tends to be focused on acquiring technical skills that are of particular value for economic development. . . . Politically the most significant feature of the process of acculturation within the army is that it usually provides some form of training in citizenship.[13]

Morris Janowitz speaks of the military offering training for technical and administrative skills as well as basic literacy and citizenship.[14] Lerner and Robinson note that the Turkish army became "a major agency of social change" precisely because it "spread among [Turkish rural youth] a new sense of [national] identity—and new skills and concepts as well as new machines."[15] But can the personality changes wrought by army service bring long-range changes in African states, given that universal military training does not exist? And, more fundamentally, is the functional specialization of armed services conducive to political development? The answer to this second question is a qualified "no," due to factors we shall explore subsequently.

Popular Participation and Identification

The extent of political participation can readily be altered by military regimes; the degree of popular identification with the political system is not susceptible to easy manipulation.

In the creation of political systems, perhaps the most important task is that of fostering popular identification. People who physically are members of a given political system should come to feel psychologically members of that system.[16] Analysts often presume that increased political participation indicates greater psychological identification with the system as a whole. However, as Clifford Geertz and others have noted, enhanced opportunities for participation may stimulate

> an obsessive concern with the relation of one's tribe, region, sect, or whatever to a center of power that, while growing rapidly more active, is not easily either insulated from the web of primordial attachments, as was the remote colonial regime, or assimilated to them, as are the workaday authority systems of the "little community." Thus, it is the very process of the formulation of a sovereign civil state that, among other things, stimulates sentiments of parochialism, communalism, racialism, and so on, because it introduces into society a valuable new prize over which to fight and a frightening new force with which to contend.[17]

Popular identification cannot be legislated into existence. As Morroe Berger commented in his analysis of the Egyptian military, "some realms of life cannot be directly affected in an enduring way through the machinery of government."[18] To be certain, the overthrow of an unpopular civilian dictator may bring great initial credit to the perpetrators of military intervention; installation of a genuinely popular army member as head of state similarly may result in public approbation. The vagaries of popular opinion, however, and the threat of ethnically based parochialism, make widespread popular identification with the new regime an unlikely direct product of military intervention.

Let us recapitulate the three major aspects of political development. As organizations, the armed forces of African states illustrate centralized authority and functional specialization. Seizure of control by the military may result from confusion over the source of authority within the state. There is no guarantee that the new leaders resulting from a coup d'etat will be viewed by the people as exercising *legitimate* authority—a possibility that, in Hobbes's terms, opens the way for further disputes and use of force. The growth of identification with the political system (surely a key part of legitimacy) may well depend on factors largely outside the area of direct governmental control. Despite the changes wrought in individuals by military training, there is no guarantee that the concepts of national unity purportedly brought by this training will affect, in the short run, the perceptions and behavior of most inhabitants of the state concerned. Political development, as the recent history of Turkey illustrates, requires a combination of skillful leadership, fortuitous circumstances, national unity, and political capacity unlikely to recur in most African states now under military rule.

II. Political Developments in Turkey: An Example for Africa?

At the close of World War I, the once glorious Ottoman Empire lay in ruins. The hegemony of the Sublime Porte, which had been exercised over an area extending from the Balkans through the Near East, had been effectively broken. The government of the Sultan, confronted with the ineluctable fact of military defeat, was forced to concede to the Allied Powers what little remained of its Near Eastern territories. However, the surgery performed by the victorious Allies, coupled with effective leadership by a former Army general, appear to have cured the "Sick Man of Europe."

The success of Kemal Ataturk frequently is cited as evidence that military rule can bring transformations civilian regimes were unable to bring about. "Be like Ataturk," the adage seems to run, "and our country will prosper." However, to assume that the installation of a government headed by members of the military brings about political development simply is incorrect. The experiences of Turkey under Ataturk ("Father of the Turks") point up the severe limitations of military rule unless it is linked closely with the development of effective political institutions.

Turkish forces did not distinguish themselves during World War I—with the major exceptions of the Gallipoli campaign in late 1915 and the defense of Aleppo in October 1918. In both these campaigns, Kemal was the leader; he was the sole Turkish military leader to enhance his reputation during the war.[19] The capitulation of the Sultan's government to humiliating terms (the Mudros Armistice) and subsequent occupation of Istanbul and parts of Anatolia stirred great discontent. Ataturk, under the guise of supervising the disarming and demobilization of remaining Turkish forces, established a base in northern Turkey from which he defied the subservient Istanbul government and challenged the invading Greek armies. The "Eastern Anatolian Society for the Defense of National Rights" called for defense against occupation or foreign interference; should the Sultan's government prove incapable of defending the nation's independence and integrity, a provisional government would be formed.[20] The Anatolian nationalists extended control over the civilian administration. Military success against Armenian, French, and Greek forces resulted, by 1922, in a new armistice and subsequent treaty recognizing the independence of Turkey within the 1918 armistice lines (Istanbul and Asia Minor).

It is not my purpose to summarize Turkish history since 1918.[21] However, five basic points must be made.

1. The multinational Ottoman Empire, whose leaders were torn among conflicting objectives of Pan-Ottomanism, Pan-Turanism, Islamic solidarity, or Turkish nationalism, could not readily succeed. Ataturk departed from Ottoman policy by recognizing the "necessity for shrinking the Ottoman Empire down to a small, culturally more homogeneous and politically more viable Turkey. . . .[22]

2. The military in Turkey had long played a basic role in introducing new techniques and "Western" ideas;[23] Ataturk thus continued—and expanded—trends that characterized the Ottoman military establishment.

3. Ataturk envisaged nothing less than a total transformation of Turkish society. He did not intend to intervene briefly to ameliorate a few minor ills; his objectives were nothing short of revolutionary.

4. Turkey benefited from Ataturk's relatively long rule. Continuity of leadership, combined with relative domestic tranquility

after 1923 and the absence of international pressures, eased the long-range changes with which Ataturk was concerned.

5. Despite his military background, Ataturk "civilianized" himself and his regime as rapidly as feasible. His official portrait, to take one example, showed him in white tie and tails, not military regalia, reflecting his "determination to civilianize the Turkish Republic that cut wide and deep."[24]

Ataturk's inimitable foresight rested essentially upon a clear recognition of the need for effective political institutions. His was not a self-serving military clique or junta but a government that built popular support for government by means of economic development, gradual mobilization of support as through the Houses of the People, and (of course) the charisma of Ataturk. The pragmatic Kemalist policy was founded upon populism and etatism—upon an astute combination of limited popular participation and centralized control within a gradually changing social framework. As Lerner and Robinson note, "Central political control was essential to regulate institutional innovation—so that changes would occur fast enough, but not so fast that social mobility would undermine political stability."[25] Or to quote a similar assessment by Janowitz, Turkey under Ataturk "represented the one case in which a military oligarchy under an enlightened leader made fundamental contributions to social and economic modernization. The process of modernization in Turkey was gradual and built on evolving institutions."[26] The example of Turkey thus shows that political development cannot result from a simple recognition of the needs for centralized control, differentiation and specialization of institutions, and greater political participation and identification.

III. Political Development and Institutional Capacities

To examine political change adequately, we must remain continually aware of the *capacity* of political institutions to satisfy demands.[27] Marked governmental ineffectiveness in the face of mounting popular pressures characterized most military-dominated African states before intervention occurred. Rapid social mobilization[28] results in increased and unrealistic demands upon governments. It abets, as well, centrifugal tendencies, in which "primordial sentiments" of assumed blood ties, race, language, region, religion, or custom override sentiments of "national" unity.[29] The mutiny of Ibo officers in the Mid-West Region of Nigeria, turning the Region over to Biafran control in August 1967, epitomizes the triumph of "primordial sentiments."

To measure the "capacity" of political institutions is an ambitious task. Capacity depends upon numerous interrelated variables, the combination of which obviously alters over time in all states. At the risk of oversimplification, however, I wish to consider capacity in terms of (1) "legitimate organizational strength," and (2) sufficient resources to carry out tasks.[30] The key factor is authority—an admittedly nebulous but central term in the analysis of the capacity of political institutions.

Political institutions exercise authority by exercising control that is seen as rightful, correct, or legitimate. Authority implies a recognition of common purposes and a belief in the inherent rightness of leadership of some sort. Most citizens must be willing to bridle private or personal impulses for the sake of general social objectives. Rapoport comments cogently that this organizational strength characterizes both civilian and military life: recognition of a "rule of law" in the civilian realm, discipline in the army, channel individual desires toward a common good.[31]

In many states, both Western and non-Western, nationalism provided cohesiveness and purpose. A sense of unity was enhanced by historical ties; defense against an external enemy; commonality of tradition, language, and territory; and the "intimate interconnections"[32] of state and nation. A sense of community emerged. Governments were recognized as legitimate; their authority rested not upon coercion and force of arms but upon expectations of compliance and obedience.

In the nineteenth century, one would be hard pressed to uncover traces of such nationalism in the Ottoman Empire. The Sublime Porte ruled over a territory suffering from a "refractory set of ingredients for a modern nation-state: it stretched over parts of three continents; lacked racial, religious, and linguistic homogeneity; included peoples with separate and competitive traditions and aspirations; and possessed few natural frontiers."[33] However, the vicissitudes of World War I reduced the Ottoman Empire to little more than its Anatolian heartland. The repatriation or repression of the Greek and Armenian minorities further enhanced ethnic homogeneity. Islam (though regarded by Ataturk as retrograde in its impact) provided further common bonds. Ataturk added his personal prestige and magnetism to the institutions of the Turkish Republic. The key factor in Ataturk's effort was the stimulation of nationalism. "Geographical and political nationalism was the device to which Kemal turned," Robinson comments. "There was a deliberate and sustained effort to weld together a Turkish nation."[34]

The necessity for a similar effort confronts African political leaders. It has frequently been stressed that African states are riven with differences arising from conflicting "primordial sentiments." As the subtitle of a recent book inquired, should Nigeria be considered in terms of tribes, people, or race? The fact is often expressed in the antimony of "tribalism" and "nationalism" —admittedly inadequate and possibly misleading terms to characterize complex phenomena. More appropriately, one should say that ethnic loyalties for most Africans are not yet fully complemented by identification with the country as a whole: Aizo, Fon, Bariba rather than Dahomey; Bamileke, Bassa, Bulu rather than Cameroon. Recognizing the limited extent of "national" loyalties, or, in somewhat different terms, the low level of national political integration, African political leaders have often viewed the state, or a dominant party, as the "architect" of the nation.[35] However, 'twixt concept and implementation, theory often goes astray. The inefficiency of the government and the decay of the dominant party into little more than a machine for distributing spoils[36] reduce the necessary element of institutional strength. More significantly, the process of social mobilization in ethnically heterogeneous African states often reinforces particularistic tendencies.[37] "The African nationalist still has before him almost the entire task of creating the nations in whose name he professes to speak," Emerson comments.[38] Nigeria has been established—but not Nigerians. Accordingly, to build political authority upon the shallow foundations of Nigerian, Dahomeyan, or any other "nationalism" in Africa runs serious risks.

Obviously, it would be preposterous to assert that nationalism as a factor underlying political authority in African states cannot be enhanced. The major question, rather, is whether African governments *under military auspices* will have greater likelihood than their civilian predecessors in developing nationalism as a basic support for political authority. Are the problems of a type that the particular strengths of army-dominated governments can better resolve? To provide tentative answers to this question, let us return briefly to the example of Turkey, and then analyze the actions of African military regimes since their assumption of control.

The Turkish example seems inapplicable to contemporary Africa in two respects. First, Turkey, after the establishment of the Republic, manifested a degree of ethnic, linguistic, and religious homogeneity more extensive than most African countries. The multinational Ottoman Empire suffered from social cleavages that negated the efforts of the Young Turks. Their vision of a modernized empire caused considerable upheavals among both Christians and Muslims and eventually stimulated Arab nationalism—all of which defeated the vision.[39] Reduction of the empire to its Turkish heartland preceded the effective implementation of nationalism.[40] Arguing by analogy (though hardly the most defensible form of argument!), the development of nationalism as a foundation for political authority in Africa would seem to require building upon existing ethnic foundations rather than creating a divided and weak multiethnic state. Second, to call Ataturk's regime "military" seriously misrepresents the facts. His policies of civilianization, previously alluded to, established a clearer division between military and civilian aspects of government than existed in the Ottoman period.

Granted then, that the ethnic divisions of African states impede the growth of political legitimacy based upon nationalism,[41] are there other bases on which military-dominated governments might seek to achieve legitimacy? The jubilation that accompanied the overthrow of Nkrumah and other unpopular politicians would appear to indicate widespread approbation for various coups. In the words of Roger Murray, the leaders of intervention profited from "popular exhaustion and tacit or active support for the dismantling of sclerosed political institutions."[42] By removing venal and autocratic rulers, whether by incarceration, assassination, or exile, the military rulers emphasized the break with the past. Political parties were banned, on the basis of the corruption and authoritarianism of their previous leaders. One may ask, however, whether this distaste for politicians and support for the military will persist for more than a few months.

No groups of rulers can long base their claim for legitimacy on acrimony and memory of previous injustices. If the new military governors of African countries intend to remain in power for an extended period (what constitutes "extended" obviously varies from state to state), they must seek to build the capacity of governing institutions, in terms both of legitimate organizational strength and skillful use of resources. There are three interlocking ways that African military rulers might seek to achieve legitimacy: (1) gain or create social and political basis of support; (2) avoid recourse to excessive use of force; (3) build effectiveness over an extended period. Given the contemporary characteristics of both African states and their military establishments, none of these means will likely bring about legitimacy. "Prolonged effectiveness over a number of generations," Lipset notes, "may give legitimacy to a political system. In the modern world, such effectiveness means primarily constant economic development."[43] This pathway to legitimacy is not feasible for governing African juntas. Both the uncertainty of economic development in Africa and the timespan emphasized by Lipset imply that short-run effectiveness will not be translated into widespread support for military governments.

The tendency to resort to violence exists in all military-dominated governments[44]—after all, how else did the military gain control? As Parsons and Zolberg have demonstrated, the use of force quickly erodes the basis of legitimate authority on which governments are based.[45] Devised to use force in the most efficient manner, armies have a natural proclivity to turn to violence rather than palaver, to repression rather than compromise. Bargaining and compromise are not familiar skills to military leaders. In scorning politicians, they may scorn the techniques well known to politicians, such as the artful ethnic balance. Military life does not necessarily provide relevant, simple formulae for a viable political system; army leaders must learn politics through experience.[46] Let me cite one example from the ill-fated Nigerian regime of General Aguiyi-Ironsi. The abortive attempt to implement Decree No. 5 in May 1966, "to remove the last vestiges of the intense regionalism of the recent past, and to produce that cohesion in the governmental structure which is so necessary in achieving and maintaining the paramount objective of the National Military Government, and indeed of every true Nigerian, namely, national unity," in the words of General Ironsi, bore witness to haste, a lack of political realism, an assumption that the centralized structure of the army could be transplanted simply into a complex government. Elemental prudence would dictate, accordingly, that army leaders minimize resort to their specialization—effective utilization of violence—for the methods of compromise and negotiation developed by politicians. Successful translation of this observation into prudent action, however, is unlikely.

The most feasible means of building legitimacy rests in careful cultivation of a strong basis of popular support. Military rulers, like their civilian predecessors, must construct a coalition of supporting groups, create political parties with widespread appeal, develop strength among significant parts of the population. Many coup leaders have sought legitimacy by naming civilian advisory councils, often in response to mounting discontent with policies imposed by the military governments. In Ghana, for example, a civilian "Political Committee" was named on June 30, 1966, four months after Nkrumah's ouster, when austerity policies were causing increasing unrest. The Gowan regime in Nigeria created a civilian advisory council in June 1967; its creation was intended to smooth the creation of twelve states within a shifting federal framework.

The role of civilian councils has not been significant. In Dahomey, the civilian Committee for National Renovation was dissolved in June 1967, when its members, in the eyes of Soglo, seemed more concerned with revivification of regional splits and divisive political tendencies than with national unity. In Ghana and Nigeria, a great deal of pressure was necessary to bring civilians into the cabinet, and this sharing of power resulted from emergency conditions (e.g., devaluation of the cedi, Biafran secession, and the creation of new states) that necessitated enhanced support. Civilian advisory councils should not be regarded as nuclei of future governments but as tactical allies of the military rulers, essential for increasing their authority.

Of necessity, army-dominated African governments have worked closely with civil servants.[47] Military governments, having eschewed the politicization of civil services that often had been enforced under single-party regimes, have enhanced the role of civil servants. Indeed, it is tempting to assert that the similar technocratic orientation and organizational hierarchy of army officers and civil servants bring both groups into natural alliance. Whether an alliance of selected civilians, civil servants, and the military results in a political elite that can achieve legitimacy must be answered on the basis of individual instances—but, in keeping with the generally pessimistic tone of this paper, I suggest that the three-way coalition will not be cohesive and that it will lack the "identific" element of sympathy from the population.[48]

Military leaders have other options. Rather than draw upon

the ranks of former politicians and highly placed civil servants, they can create new political parties under their direct auspices to foster legitimacy. In other words, instead of drawing upon the support given to politicians participating in an elite coalition, army officers may seek their own foundations of popular support. General Mobutu, for example, announced in April 1967 his intention of creating and leading the "Mouvement populaire de la Révolution," intended to organize the popular masses for education, instruction, and training. But the prospects for the rapid creation of a strong party structure are negligible. In the face of widespread apathy, attempts to build legitimacy through persuasion, party activity, and economic development involve huge expenditures of effort without significant short-run impact. The series of failures to create a broadly based political movement in the United Arab Republic, despite intensive efforts by President Nasser, and the indifferent success of Ataturk himself in developing part support among the peasantry, clearly indicate the limitations on building legitimacy through party auspices.[49]

To rule effectively, once having achieved power, military leaders must develop political organizations of civilian types or work out viable relations with civilian political groups.[50] There are no other paths toward political development, as defined in this discussion. Military governments will confront the same problems of ethnicity and economic stagnation that confronted civilian governments. The centralized authority and functional specialization of armies may promote intervention, but they do not enhance the possibilities for effective governance based upon legitimate governmental strength. The military rulers of Africa often lack bargaining skills. Their impatience with politicians—their "politics of wanting to be above politics"[51]—leads them to distrust the slow, difficult process of coalition-building. Having banned political parties, and likely lacking the abilities to build parties afresh, the African military may well not promote political development. They cannot build legitimate political institutions through the use of force; they cannot transplant their organizational style to the civilian realm; they are not equipped to bring stability, modernization, or political participation; they lack sufficient flexibility and innovativeness to govern effectively.[52] My prognosis—which may seem overly bleak—is for political decay, not political development, in African states falling under military control.

NOTES

1. David E. Apter, *The Politics of Modernization* (Chicago: University of Chicago Press, 1965); James S. Coleman, ed., *Education and Political Development* (Princeton, N.J.: Princeton University Press, 1965); John H. Kautsky, ed., *Political Change in Underdeveloped Countries: Nationalism and Communism* (New York: Wiley, 1962); Joseph La Palombara, ed., *Bureaucracy and Political Development* (Princeton, N.J.: Princeton University Press, 1963); Joseph La Palombara and Myron Weiner, eds., *Political Parties and Political Development* (Princeton, N.J.: Princeton University Press, 1966); Marion J. Levy, Jr., *Modernization and the Structure of Societies* (Princeton, N.J.: Princeton University Press, 1966); A. F. K. Organski, *The Stages of Political Development* (New York: Knopf, 1965); Lucian W. Pye, *Aspects of Political Development* (Boston: Little, Brown, 1965); Lucian W. Pye, ed., *Communications and Political Development* (Princeton, N.J.: Princeton University Press, 1963); Lucian W. Pye and Sidney Verba, eds., *Political Culture and Political Development* (Princeton, N.J.: Princeton University Press, 1965); Dankwart A. Rustow and Robert E. Ward, eds., *Political Modernization in Japan and Turkey* (Princeton, N.J.: Princeton University Press, 1964); Edward Shils, *Political Development in the New States* (The Hague: Mouton, 1962); K. H. Silvert, ed., *Expectant Peoples: Nationalism and Development* (New York: Random House, 1963); Claude E. Welch, Jr., ed., *Political Modernization: A Reader in Comparative Political Change* (Belmont, California: Wadsworth, 1967).
2. Welch, *Political Modernization,* p. 7.
3. S. E. Finer, *The Man on Horseback: The Role of the Military in Politics* (London: Pall Mall, 1962).
4. Samuel P. Huntington, "Political Modernization: America vs. Europe," *World Politics* XVIII, 3 (April 1966), 378.
5. Carl J. Friedrich, ed., *Authority* (Cambridge, Mass.: Harvard University Press), pp. 13–58.
6. Carl J. Friedrich, *Man and His Government* (New York: McGraw-Hill, 1963), p. 244. Robert Dahl defines authority as the influence of a leader clothed with legitimacy; legitimacy is the belief that structures, procedures, acts, decisions, policies, officials, or leaders

of government possess "the quality of 'rightness,' propriety, or moral goodness and should be accepted because of this quality." Robert A. Dahl, *Modern Political Analysis* (Englewood Cliffs, N. J.: Prentice Hall, 1963), p. 19.

7. Dahl, *Modern Political Analysis,* pp. 31–32.
8. Karl W. Deutsch et al., "Political Community and the North Atlantic Area," in *International Political Communities: An Anthology* (Garden City, N.Y.: Doubleday, 1966), pp. 1–91. Deutsch and his associates list nine essential conditions for the creation of an "amalgamated security-community": (1) mutual compatibility of main values; (2) a distinctive way of life; (3) expectations of stronger economic ties or gains; (4) a marked increase in political and administrative capabilities of at least some participating units; (5) superior economic growth on the part of at least some participating units; (6) unbroken links of social communication, both geographically between territories and sociologically between different social strata; (7) a broadening of the political elite; (8) mobility of persons, at least among the politically relevant strata; and (9) a multiplicity of ranges of communication and transaction. The authors note that three other conditions may be essential: (10) a compensation of flows of communications and transactions; (11) a not too infrequent interchange of group roles; and (12) considerable mutual predictability of behavior. *(International Political Communities,* pp. 37–38.)
9. C. S. Whitaker, Jr., unpublished paper presented at the annual meeting of the African Studies Association, 1965. Chief Awolowo, in his recent book, notes, "historical evidences of an irrefutable nature have shown firstly, that you can unite but can never succeed in unifying peoples whom language has set distinctly apart from one another; and secondly, that the more educated a linguistic group becomes, the stronger it waxes in its bid for political self-determination and autonomy, unless it happens to be the dominant group." Obafemi Awolowo, *Thoughts on Nigerian Constitution* (Ibadan: Oxford University Press, 1966), pp. 54–55.
10. Guy J. Pauker, "Southeast Asia as a Problem Area in the Next Decade," *World Politics* XI, 3 (April 1960), 339.
11. Fred W. Riggs, *Administration in Developing Countries: The Theory of Prismatic Society* (Boston: Houghton Mifflin Company, 1964), p. 23.
12. Lucian Pye notes, for example, that "modern armies are essentially industrial-type entities." "Armies in the Process of Political Modernization," in J. J. Johnson, ed., *The Role of the Military in Underdeveloped Countries* (Princeton, N.J.: Princeton University Press, 1962), p. 76.
13. *Ibid.,* pp. 80–83.
14. Morris Janowitz, "The Military in the Political Development of New Nations," in Wilson C. McWilliams, *Garrisons and Government: Politics and the Military in New States* (San Francisco: Chandler, 1967), p. 74; cf. Pye, "Armies in the Process of Political Modernization," pp. 76 ff.

15. Daniel Lerner and Richard D. Robinson, "Swords and Ploughshares: The Turkish Army as a Modernizing Force." See Chapter 4 above.
16. Sidney Verba, "Comparative Political Culture," in Lucian W. Pye and Sidney Verba, eds., *Political Culture and Political Development* (Princeton, N.J.: Princeton University Press, 1966), p. 529.
17. Clifford Geertz, "The Integrative Revolution: Primordial Sentiments and Civic Politics in the New States," in Welch, *Political Modernization,* p. 177.
18. Morroe Berger, "The Military Elite and Social Change: Egypt since Napoleon," in McWilliams, *Garrisons and Government,* p. 225.
19. Richard D. Robinson, *The First Turkish Republic: A Case Study in National Development* (Cambridge, Mass.: Harvard University Press, 1963), p. 24.
20. Robinson, *The First Turkish Republic,* p. 68.
21. Kemal H. Karpat, *Turkey's Politics: The Transition to a Multi-Party System* (Princeton, N.J.: Princeton University Press, 1959); Richard H. Pfaff, "Disengagement from Traditionalism in Turkey and Iran," *Western Political Quarterly* XVI, 1 (March 1963), 79–98: Robinson, *The First Turkish Republic;* Ward and Rustow, eds., *Political Modernization in Turkey and Japan.*
22. Robinson, *The First Turkish Republic,* p. 22.
23. Dankwart A. Rustow, "The Military," in Ward and Rustow, *Political Modernization in Japan and Turkey,* pp. 352–388.
24. Lerner and Robinson, "Swords and Ploughshares," p. 118 above.
25. *Ibid.,* p. 122.
26. Morris Janowitz, "The Military in the Political Development of New Nations," in McWilliams, *Garrisons and Government,* p. 271.
27. I am arguing here along lines analogous to Professor Spiro's concern with the "development of politics," namely that "expansion of the capacity of the political system to process a heavier volume of problems, to discuss a greater range of issues, to forge new goals as old ones are being approached, and to make more people conscious of the possibilities of change-through-politics." Herbert J. Spiro, "The Military in Sub-Saharan Africa," in McWilliams, *Garrisons and Government,* p. 271.
28. Deutsch, "Social Mobilization and Political Development" (information fn. 37).
29. Geertz, "The Integrative Revolution," in Welch, *Political Modernization,* pp. 171–172.
30. Huntington, in his provocative critique of theories of political development, focuses upon institutional capacities in terms of scope of support and level of institutionalization, the latter being measured by the adaptability, complexity, autonomy, and coherence of its organizations and procedures. Samuel P. Huntington, "Political Development and Political Decay," see Chapter 6 above.
31. David C. Rapoport, "A Comparative Theory of Military and Political Types," in Samuel P. Huntington, *Changing Patterns of Military Politics* (New York: The Free Press, 1963), p. 79.

32. Rupert Emerson, *From Empire to Nation: The Rise to Self-Assertion of Asian and African Peoples* (Cambridge, Mass.: Harvard University Press, 1960), p. 104.
33. Ward and Rustow, *Political Modernization in Japan and Turkey,* p. 438.
34. Robinson, *The First Turkish Republic,* p. 84.
35. Leopold Sedar Senghor, *Nation et Voie Africaine du Socialisme* (Paris: Présence Africaine, 1960), p. 24.
36. Austin notes, for example, that the C.P.P. turned into a massive propaganda machine; the privileges of a large number of beneficiaries depended on its survival. Dennis Austin, *Politics in Ghana 1946–60* (London: Oxford University Press for the Royal Institute of International Affairs, 1964), p. 418.
37. Karl W. Deutsch, "Social Mobilization and Political Development," *American Political Science Review* LV, 3 (September 1961), 493–514; Geertz, "The Integrative Revolution," in Welch, *Political Modernization,* pp. 167–187.
38. Emerson, *From Empire to Nation,* p. 94.
39. Karpat, *Turkey's Politics,* p. 24.
40. Ward and Rustow, *Political Modernization in Japan and Turkey,* p. 438.
41. I am distinguishing here between nationalism and anticolonialism. This distinction has been criticized by some writers on African affairs and neglected by many others. However, the contrast between unity based on opposition to an imposed, foreign regime and unity based on a sense of shared participation within a historic entity must not be forgotten.
42. Roger Murray, "Militarism in Africa," *New Left Review,* No. 38, p. 57.
43. Seymour Martin Lipset, *Political Man: The Social Bases of Politics* (Garden City, N.Y.: Doubleday, 1963), p. 70.
44. H. Daalder, *The Role of the Military in the Emerging Countries* (The Hague: Mouton, 1962), quoted in McWilliams, *Garrisons and Government,* p. 26.
45. Talcott Parsons, "Some Reflections on the Place of Force in Social Process," in Harry Eckstein, ed., *Internal War: Problems and Approaches* (New York: The Free Press, 1963), p. 44; Aristide R. Zolberg, "The Structure of Political Conflict in the New States of Tropical Africa," unpublished paper delivered at the 1966 annual meeting of the American Political Science Association, pp. 13–14.
46. Morris Janowitz, *The Military in the Political Development of New Nations: An Essay in Comparative Analysis* (Chicago: University of Chicago Press, 1964), pp. 104–105.
47. Chief Awolowo has expressed this affinity strongly. To quote from his recent *Thoughts on Nigerian Constitution:* "It has been said that governments may come, and governments may go, but the Civil Service remains forever. It is a most merciful thing indeed that the Civil Service remains largely unaffected by the vicissitudes of politics. What a tragedy, for instance, it would have been of the recent reverses in the fortunes of politicians had hit the civil servants more or less equally! The smooth take-over by the army

would have been impossible. There would have been a chaotic collapse of public administration, with harsh and injurious effects on the welfare of people." (P. 73.)

48. Claude Ake, "Political Integration and Political Stability," *World Politics* XIX, 3 (April 1967), 491.
49. P. J. Vatikiotis, *The Egyptian Army in Politics: Pattern for New Nations?* (Bloomington, Ind.: Indiana University Press, 1961).
50. Janowitz, *The Military*, p. 1.
51. *Ibid.*, p. 65.
52. Spiro, "The Military in Sub-Saharan Africa," in McWilliams, *Garrisons and Government*, p. 266.

Index

Academic biases, regarding military, 66
Acculturation within army, 217
Adaptability, 178; and Chinese Communist Party, 168–69; in France, 173; generational age, 168; and incubation, 166–67, 198; as organizational characteristic, 167; and political system, 172–73; problems of, 171
Africa: and civil-military relations, 10; coalition, 227; effectiveness of military, 18; intervention of military, 213; and legitimacy of military, 225–26; military intervention in politics, 19; and military modernizer, 3; and military status, 6; and national liberation movement, 150; and nationalist party, 202; and nationalism, 223–24; and organizational character of army, 216; and political authority, 224; and political capacity, 24; and political decay, 229; and political participation, 179; and professional armies, 216; and social mobilization, 222, 224
America: aid, 105, 129, 137; antimilitarism, 7; and armed forces, 65–66; denegration of military, 42; faith in, 183; importance of party, 205; and international affairs, 46; and military aid, 25, 27; military, attitude toward, 7–8, 46; and military regimes, 7; and opposition to coups, 95–96; and policy in Turkey, 139; and political development, 204–6; and social scientists' attitude toward

military, 43; study of military, 42; and technical training of Turks, 129–30, 132
Anti-military organizations in U.S., 58
Apathy, 59
APRA movement, 91, 92, 94
Arévalo, Juan José, 93
Argentina: *colorados*, 91; coups and countercoups, 184; disorganization in, 184; industrial expansion in, 10; and military control, 90; and rigged elections, 94
Aristotle, 173, 191
Armed forces, 4, 44, 48, 50; assistance to Korea Program, 113; and authoritarian societies, 54; and differentiation, 216; characteristics of, 61; and civilian concerns, 73; and class membership, 63; economic aspects, 50–52; and the elites, 68–69; for external use, 45–46; and feudal societies, 45; as focus of government, 53–54; and general government, 61; in Germany, 54; internal control of, 61; internal use of, 45–46; and international affairs, 48; isolation of, 48–50, 62–65; maintenance in U.S., 42, 47–48; and modernization, 65–73; objectives, 53; particularism, 63; planning of, 44–45; political aspects of, 52–55; and radicalism, 70–71; regimentation, 50; reliance upon, 155; in Russia, 54; setting of, 44–45; and social mobility, 68–71; solidarities in, 50; and social engineering, 44–45; specialization, 52; traditional thinking of, 52; *see also* Army and Military
Army: administrative role, 14; attitude toward, 58; bureaucratic aspects of, 57; in communist China, 102; composition of, 151; concern about, 2; and contests of skill, 57; and control of personnel, 57–62; and defense, 45; democratic tendency, 154; and discipline, 178; and efficiency, 11–12; as guiding principle, 156; and hierarchy, 155; as ladder of social mobility, 12–13; and Middle East, 19; and modernizing influences in Latin America, 11; as modernizing force, 15; as national force, 150; and national liberation movement, 149; and national revolution forces, 152; neutrality of, 150; as offense, 45; officers of, 129; and political orientation, 55; reactionary tendencies in, 154; and reconstruction, 212; and revolutionary change, 10; role in revolution, 22; social background of members, 63, 64; and socialization, 58; and tribal complexity, 18; in Turkey, 25, 118, 119, 126–27, 143–44; as vanguard, 22; *see also* Armed forces; Military
Asia, 3, 10, 13, 28, 103, 106
Ataturk model: and civilian institutions, 20; and economic development, 124; and etatism, 124; and nationalism, 223; and political development, 14–17; and populism, 124; relationship with Ismet Inonu, 118
Authority: centralization of, 214–16; and control, 215; definition of, 215; emphasis on, 72; institutionalized, 23; and military, 14; and rank, 57; source of, 215; traditional, 56, 215
Autonomy, and coherence, 177; and corruption, 181; importance of, 174; measure of, 178; and political institutions, 174–75; and political procedures, 175; and subordination, 174–76
Baghdad Pact, and nonmilitary construction, 138
Balkan Pact, 138
Belaúnde, Fernando, 92
Berger, Morroe, 218
Biafra and control of Nigeria, 222
Bolshevik: concept of party, 204; organization size, 202
Bosch, Juan, 93–94
Bourgeoisie and national liberation, 151

Branco, Marshal Castelo, 89-91
Brazil: and conflict between military and existing authority, 91; industrial expansion of, 10; *linha dura* in, 91; and Joao Goulart, 90
British tradition of civil supremacy, 107
Budget for national defense, 126
Bureaucracy: aspects in armed forces, 57; in Latin American countries, 185; as inevitable, 12
Burma, 6, 113
Burmese Socialist Party Program, 156
Bushido, as communication in Japan, 112

Caesarism, 10
Cakmak, Marshall Ferzi, 119
Caretaker rule, 108
Caste associations and Indian politics, 194–95
Caste lines, 128–29
Centralization: of authority, 214–16; of power, 214
CENTRO, *see* Baghdad Pact
Chain of command and communication, 214
Chang regime, 105
Charismatic leaders, 182, 206
China: military, 112; modernization and conflict in, 110
Chinese Communist Party, 168–69; on political tasks, 205
Chinese Red Army, 113
Civic action, 25, 26, 27; and counterinsurgency, 26
Civilian control of military in Turkey, 126
Civilian councils, 227
Civilian institutions: Ataturk model of, 20; as modernizers, 21
Civil-military: compatibility, 140; interdependence, 51; relations 10; relations after World War II, 1
Civilian leadership, and control in Turkey, 122, 145–46
Civilian political groups, 24
Civilian government, 106
Civilian rule, 1, 14–15
Civilian supremacy, 124; in Turkey, 15–16, 120, 121, 141, 143; subversion of, 145
Civil innovation, 134–35
Civil War, in U.S., 46
Class membership, *see* Armed forces
Coercion, 12; stability by, 140, 54–55
Coherence, 176–79
Cold War decade, 128
Command, emphasis on, 12, 54–55
Committee of National Unity, 143
Communication: external environment and, 109; intrasocial network and, 109, 110, 111; and the military, 13; and mobilization, 180, 195; and modernity, 180, 195; symbolic, 112; with West, 109
Communism: and literacy, 180; and military regimes, 9; national armies as counterbalance, 8; and political participation, 204
Communist China, army and dictatorship in, 102
Communist states, military training in, 110
Complexity-simplicity, 171–74
Compulsory military service, 132
Congressional hearings on military assistance, 9
Conscription, 67–68
Conservatism and the military, 17
Constitutionality in Latin America, 81, 94
Corruption: and autonomy, 181; and coherence, 181; and division in military, 123; and economic change, 181; and social change, 181; increases in, 182; in recruitment, 68
Coups d'etat, *see* Coups
Coups: American opposition to, 95–96; conditions for, 82; and conservative regime, 5; and control, 214; deterrents to, 27, 96; dynamics of, 4, 86–94; and economy, 82–83; and elections, 88; external factors of, 83–84; and

factionalism, 20; influence of, 175; and intervention, 3; in Latin America, 81–83; and military control, 90; and military officers, 86; and military rule, 4; and military support, 87–88; in Peru, 92; and political change, 6, 83–85; and political institutionalization, 181; and political order, 24, 206; and political orientation, 86; and political problems, 6; and political stability, 182; and political systems, 86; precipitating factors of, 95; and susceptibility of, 104; and swing man, 89; variables of, 104
Cyclical factors, and dictatorship, 5, 97–98

Dahl, Robert, 106
Defense and the military, 42, 45
Defense Service Resettlement Program in Burma, 113
Degeneration, political, 190–92
de la Torre, Victor Raúl Haya, 91
Democracy: erosion of, 163; and literacy, 180; and the military, 14; and military regimes, 9; in Turkey, 140
Democratic tendency of Army, 154
Democratization, 159
de Tocqueville, Alexis de, 157, 204
Deutsch, Carl, 160
Development: criteria of, 162; as a quality, 161; Western concept of, 161
Deviant behavior, 60
Dictatorship, 5, 23, 97–98
Differentiation and specialization, 216–17
Discipline: in Africa, 18; and development, 178
Dominican Revolutionary Party, 93–94
Draper report, 9, 26–27

Eastern Anatolian Society for the Defense of National Rights, 220
Economic: aspects of armed forces, 50–52; aspects of development, 26; change and corruption, 181; deterioration, 4; development in Turkey, 127; instability, 180; mobilization, 196
Education: in armed forces, 43, 71–72; in Turkey, 135–36, 139
Effendi, Abdurrahim, 125
Egalitarianism, 55, 134
Elections: and coups, 88; in South Korea, 108; in South Viet Nam, 108
Engels, F., 151
England, and policy of enclosure, 16
Etatism and populism, 124
Evolution: of armed forces, 4; of new states, 191

Factionalism, 20
Feudal societies, 45
Force: and political stability, 54; reliance on by civilian authority, 19; resort to by military, 226; use of by military, 52–53
Foreign policy, 123
French political adaptation, 173
Functional: adaptability, 171; specialization, 217

Gazi, as military commander, 118
Geertz, Clifford, 218
Gemeinschaft-Gesellschaft and military leaders, 8
Gendarmerie as policing force, 126
Germany and armed force as government, 54
Glick, Edward, 26
Godoy, Gen. Richard Perez, 92
Governing: emphasis on, 54–55; and the military, 54; and professional soldiers, 12
Gowan, in Nigeria, 227
Graft in recruitment, 68
Great Britain, 46, 187
Gursel, Gen. Cemal, 143–45

Halpern, Manfred, 17, 18, 21
Harvard, 166
Hee, Gen. Park Chung, 106
Helling, George, and Turkish soldier, 135
Hierarchy: in armed forces, 55; and tradition, 56

Highway development program in Turkey, 137
House of the People, 221
Huntington, Samuel P., 5, 21, 23, 214

Ibo officers, 222
Incubation and adaptation, 198
India, 184, 202
Indonesia, 102, 105, 107–8, 112–13
Inonu, Ismet, 118, 144
Instability and mobilization, 192
Institutions: adaptability of, 166–67; building, 192–93; creation of, 197–200; decay of, 181; definition of, 166; development of, 194; and government organizations, 187; and intervention, 6; and modernization, 181; and political organizations, 166, 169; and public interest, 188; strategies of, 192–206
Insurgency: and civic action, 26; in Latin America, 85
International affairs: and armed forces, 48; and the U.S., 46
Isolation: and age of members of armed forces, 49; and modernization problems, 72; and morale, 60; and partisan politics, 126; and readiness, 62

JAMMAT (Joint American Military Mission for Aid to Turkey), 131
Janowitz, Morris, 86
Japan, 54, 61, 69, 70, 112, 193
Johnson, John J., 185
Johnson, Lyndon B., 166
Judiciary, independence of, 174
Juntas: and political order, 24; problems of, 23; and swing man, 93

Kayeri, and law, 144
Kemalist policy, 124, 221
Khan, Gen. Ayub, 102, 106
Khan, Liaquat Ali, 105
Khan, Gen. Mohammed Ayub, 107
Khrushchev, N., 189
Kikuyu, 18
Korea, 6, 24; *see also* North Korea; South Korea
Korean Military Revolution of 1961, 111

Ladder of social mobility, 12–13
Latin America: changes in, 86–87; and the coup d'etat, 80, 81–83; and de facto government, 80; and growth of professionalism, 80; and identity crisis, 181; and insurrections, 85; and integration of military, 10; and military bureaucracies, 185; and military intervention, 6, 80, 182; and modernizing abilities, 11; and political development, 80; and political institutions, 183; and role of military, 2, 9–10, 28; and social mobilization, 80; and U.S. influence, 80; and voting, 181
Latin Americanists, 79
Legitimacy of government, 189; and the party, 201; and popular support, 226
Lenin, V. I., 22, 150, 205
Lerner, Daniel, 15, 160, 180
Liberation movement, 155
Literacy, 180
Lieuwen, Edwin, 97, 104
Linha dura, 91
Logistics, 50
Lopez, Gen. Lindley, 92

Macaulay, on corrupt society, 183
Machiavelli, 191, 199
Maginot Line attitude, 52
Magsaysay, President, 113
Maintenance crisis, 131
MAP (Military Assistance Program), 25–26
Marshall, John, 188
Marxism, teaching of, 111
Marxist, and state, 174
Marxoid group, 91
Mehta, Ashoka, 168
Meiji restoration, 193
Menderes regime, 144
Middle East, 17, 19, 21
Military, 2, 6, 9, 11, 20, 42, 43, 50, 54, 119; as administrators, 54;

and age classification, 49; and American aid, 25; analysis of, 3; assistance, 26, 27; and the budget, 16; and bureaucracy, 185; characteristics of, 15; and civilians, 19, 22; cohesion, 6; and control, 199–200, 229; and coups, 2, 87–88, 90; and de facto government, 80; and democracy, 9, 14; and developing countries, 2, 21; and dominance, 103, 106; and economic reform, 16; and education, 43, 71–72; essays on, 3; and external environment, 109; and factionalism, 20; and image, 106; and intervention, 3, 27, 84–85, 91–92, 215; and isolation, 49; as middle class, 10; as mobilizing force, 20; and modernization, 2, 3, 5, 7–11, 13, 66–67; and national unity, 218–19; and organizational structure, 1, 15, 17, 20, 137; and political power, 126, 142, 214–19; and politics, 10, 14, 20, 103–4, 108; as a profession, 86; and quasi-civilian rule, 108; and restriction of growth, 17; and revolution, 22, 58–59; role of, 1, 14, 27, 42; as ruling group, 3, 4, 7, 9, 19, 21, 123, 219, 226–27, 228; and social change, 25, 64; and social revolutions, 59; and social structures, 60–61; study of, 3, 42; and takeover, 24, 61, 145; and use of force, 52–53, 226; *see also* Armed forces
Military Aid Mission, 26,131
Minister of National Defense, 137
Mobility: and control, 69; as individual in Turkey, 122; and intellectual qualities, 129; and military membership, 69
Mobilization: and economic development, 196; and institutionalization, 179–90; and modernization, 23, 159; and participation, 165; and politics, 194, 196; as socialization, 113, 183; selective, 205; slowing of, 194–97
Mobutu, Gen., 228
Modernization: characteristics of, 162; definition of, 12; and economic development, 141; and isolation, 49, 72; objections to, 67; and political development, 158–65; and technology, 71; and urbanization, 164; as weakening force, 197–98
Morale, 59, 60, 137
Mundros Armistice, 220
Multiparty systems, 204
Mutual Security Act, 26
Myon, Chang, 107

Nation building, 193
National disintegration, 164
National force, 150, 159
National reconstruction, 212
National revolution, 152
National War College, 110–11
Nationalism: cohesiveness, 222; importance of, 24; and integration, 180; and political disintegration, 180; reality of, 163–64
Nationalization, 223–24
Needler, Martin, 4
Ne Win, Gen., 195
Nigeria, 18, 215, 222
Nkrumah, K., 182
North Korea, 110
No-party systems, 204
Nun, José, 10

Odría, Gen. Manuel, 91, 92
Offense, army as, 45
Officer: attitudes, 87; as civil servant, 73; membership, 64; and hostility, 15, 152; as modernizers, 23; as patriots, 155; and political orientation, 86; recruitment of, 11; and Southeast Asia, 8; in Turkey, 129, 133; and waiting game, 154
Opportunists, 154
Organization: building, 193; effectiveness, 203
Ottoman Empire, 124, 128, 177, 219

Palace revolutions, 58, 60
Parsonian pattern, 159
Pakistan, 105
Particularism: in armed forces, 63;

and political institutions, 182; and recruitment, 56; and universalism, 159
Party: Bolshevik concept of, 204; formation of, 205; importance of, 205; institutionalization of, 23; and political stability, 205; primacy of, 200–4; role of, 200
Pauker, Guy, 8, 215
People's party, 141
Personnel: and radicalism, 70; selection of, 133
Peru, coups in, 91–92
Philippines, 103, 112–13
Plato, 173, 191
Police force, and internal use, 47
Policy implications, 24–28
Political: changes, 163; control, 94; decay, 164; development, 158, 164, 204–6; instability, 8; and institutionalization, 165–79, 221–29; institutions, 174, 182, 186–90, 192, 198–99, 200, 201; orientation, 55; organization, 155; 156, 166; participation, 180–81; party, 23; procedures, 175; stability, 48; systems, 185, 203; vulnerability, 174
Politics: and the military, 14; concentration principle of, 202
Popular participation, 217–19
Power: application of, 61–62; as focus of action, 55; as institution, 199; irresponsible, 17; seizure of, 6
Prada, Vargas, 92
Prialé, Ramiro, 91
Professional armies: in Africa, 216; in government, 12; U.S. attitude toward, 42–43
Public interest: and differentiation, 48; in Soviet Union, 187; in the Supreme Court, 187–88
Pye, Lucien, 7–8, 14, 158, 217

Radicalism in armed forces, 70–71
Rank and authority, 57
Rapoport, David, 177, 192, 222
Reactionary tendencies in Army, 154
Reform, 28
Regimentation, 50
Report of the President's Committee to Study the U.S. Military Assistance Program, *see* Draper Report
Resources, dispersion of, 202
Revolution: and political parties, 153; of rising expectations, 134; of rising frustrations, 140
Revolutionary change: and the armed forces, 10; and democracy, 5
Riggs, Fred, 17, 216
Rigged elections: in Argentina, 94; in Turkey, 144
Rojas, Admiral Isaac, 91
Russia and armed forces, 54
Rustow, D. A., 127–28

Samurai and loyalty, 193
Scholars and the military, 42
Secret police, techniques of, 60
Secular: authority, 215; education, 136–37
Secularism and armed forces, 50
Social background of army members, 63–64
Social mobility, 68–71
Socialization, in armed forces, 58
Soldiers, as family servants, 129; *see also* Army; Armed forces; Military
South Korea: and increase in military, 105; and military education, 110; and military rule, 103; and military training, 109; and Syngman Rhee, 104–5
Specialization in armed forces, 52
Stirling, Paul, 132
Studies on military, 104
Subordination, 174–76
Supreme Council of National Defense, 119
Szule, Tad, 97

Technology, 52–53, 67, 72
Thomas, L. V., 124
Townsend Movement, 170
Traditional thinking of armed forces, 52
Training and the military, 13

Tribal complexity, 18
Truman Doctrine, 129
Tse-tung, Mao, 102
Turkey: American policy in, 139; American and European aid to, 137; army and government in, 118, 119; and Ataturk, 219, 223; inclusion in Atlantic Pact, 130; and military budget, 16; and military skills, 133; and secular intelligentsia, 122–23; technology in, 129–31; and size of army, 126–27
Urbanization: and education, 111; in Turkey, 135

Vargas, Getúlio, 91
Voting and instability, 181
WCTU (Women's Christian Temperance Union), 170
Weapons, modern, 66
Webbism, definition of, 163

For Product Safety Concerns and Information please contact our EU
representative GPSR@taylorandfrancis.com
Taylor & Francis Verlag GmbH, Kaufingerstraße 24, 80331 München, Germany

www.ingramcontent.com/pod-product-compliance
Ingram Content Group UK Ltd.
Pitfield, Milton Keynes, MK11 3LW, UK
UKHW040106200926
475702UK00011B/1264
* 9 7 8 0 2 0 2 3 6 3 0 5 9 *